Virginia Baldwin
Editor

Patent and Trademark Information: Uses and Perspectives

Patent and Trademark Information: Uses and Perspectives has been co-published simultaneously as *Science & Technology Libraries*, Volume 22, Numbers 1/2 2001.

Pre-publication REVIEWS, COMMENTARIES, EVALUATIONS . . .

"Covers an undervalued and often underused class of scientific and technical information, namely patents and trademark information, and the authors draw on their extensive experience as information professionals to make A LUCID AND IN-DEPTH PRESENTATION OF KEY RESOURCES AND INFORMATION SYSTEMS in this area. With the aid of numerous examples, the authors lead the reader in an engaging way through decision points and search paths for locating patent and trademark information from state, federal, and international sources."

Javed Mostafa, PhD
Victor H. Yngve Associate Professor
Indiana University-Bloomington

More pre-publication
REVIEWS, COMMENTARIES, EVALUATIONS . . .

"This book will surely aid information, business, and legal professionals, as well as technology and engineering researchers, in unraveling the complexities of U.S. and international patent and trademark searching. . . . EXAMINES THE TOOLS TO USE AND THE PROCESS OF CONDUCTING SEARCHES. Several value-added features that will benefit the reader are included."

June Abbas, PhD
*Assistant Professor
School of Informatics
State University of New York at Buffalo*

"This book will contribute to a better understanding of the patent and trademark research process, and is a welcome addition to the intellectual property literature. THE AUTHORS ARE KNOWLEDGEABLE AND HIGHLY TRAINED PATENT AND TRADEMARK SEARCHERS."

Claudine Arnold Jenda, MSc
*Science Librarian
and Assistant Chair
Reference and Instruction Services
Auburn University Libraries*

The Haworth Information Press
An Imprint of The Haworth Press, Inc.

Patent and Trademark Information: Uses and Perspectives

Patent and Trademark Information: Uses and Perspectives has been co-published simultaneously as *Science & Technology Libraries*, Volume 22, Numbers 1/2 2001.

Science & Technology Libraries Monographic "Separates"

Below is a list of "separates," which in serials librarianship means a special issue simultaneously published as a special journal issue or double-issue *and* as a "separate" hardbound monograph. (This is a format which we also call a "DocuSerial.")

"Separates" are published because specialized libraries or professionals may wish to purchase a specific thematic issue by itself in a format which can be separately cataloged and shelved, as opposed to purchasing the journal on an on-going basis. Faculty members may also more easily consider a "separate" for classroom adoption.

"Separates" are carefully classified separately with the major book jobbers so that the journal tie-in can be noted on new book order slips to avoid duplicate purchasing.

You may wish to visit Haworth's website at . . .

> http://www.HaworthPress.com

. . . to search our online catalog for complete tables of contents of these separates and related publications.

You may also call 1-800-HAWORTH (outside US/Canada: 607-722-5857), or Fax: 1-800-895-0582 (outside US/Canada: 607-771-0012), or e-mail at:

> docdelivery@haworthpress.com

- ***Patent and Trademark Information: Uses and Perspectives,*** edited by Virginia Baldwin, MS, MLS (Vol. 22, No. 1/2, 2001). *"A lucid and in-depth presentation of key resources and information systems in this area." (Javed Mostafa, PhD, Victor H. Yngve Associate Professor, Indiana University, Bloomington)*

- ***Information and the Professional Scientist and Engineer,*** edited by Virginia Baldwin, MS, MLS, and Julie Hallmark, PhD (Vol. 21, No. 3/4, 2001). *Covers information needs, information seeking, communication behavior, and information resources.*

- ***Information Practice in Science and Technology: Evolving Challenges and New Directions,*** edited by Mary C. Schlembach, BS, MLS, CAS (Vol. 21, No. 1/2, 2001). *Shows how libraries are addressing new challenges and changes in today's publishing market, in interdisciplinary research areas, and in online access.*

- ***Electronic Resources and Services in Sci-Tech Libraries,*** edited by Mary C. Schlembach, BS, MLS, and William H. Mischo, BA, MA (Vol. 20, No. 2/3, 2001). *Examines collection development, reference service, and information service in science and technology libraries.*

- ***Engineering Libraries: Building Collections and Delivering Services,*** edited by Thomas W. Conkling, BS, MLS, and Linda R. Musser, BS, MS (Vol. 19, No. 3/4, 2001). *"Highly useful. The range of topics is broad, from collections to user services . . . most of the authors provide extra value by focusing on points of special interest. Of value to almost all librarians or information specialists in academic or special libraries, or as a supplementary text for graduate library courses." (Susan Davis Herring, MLS, PhD, Engineering Reference Librarian, M. Louis Salmon Library, University of Alabama, Huntsville)*

- ***Electronic Expectations: Science Journals on the Web,*** by Tony Stankus, MLS (Vol. 18, No. 2/3, 1999). *Separates the hype about electronic journals from the realities that they will bring. This book provides a complete tutorial review of the literature that relates to the rise of electronic journals in the sciences and explores the many cost factors that may prevent electronic journals from becoming revolutionary in the research industry.*

- ***Digital Libraries: Philosophies, Technical Design Considerations, and Example Scenarios,*** edited by David Stern (Vol. 17, No. 3/4, 1999). *"*Digital Libraries: Philosophies, Technical Design Considerations, and Example Scenarios *targets the general librarian population and does a good job of opening eyes to the impact that digital library projects are already having in our automated libraries." (Kimberly J. Parker, MILS, Electronic Publishing & Collections Librarian, Yale University Library)*

- ***Sci/Tech Librarianship: Education and Training,*** edited by Julie Hallmark, PhD, and Ruth K. Seidman, MSLS (Vol. 17, No. 2, 1998). *"Insightful, informative, and right-on-the-mark. . . . This collection provides a much-needed view of the education of sci/tech librarians." (Michael R. Leach, AB, Director, Physics Research Library, Harvard University)*

Chemical Librarianship: Challenges and Opportunities, edited by Arleen N. Somerville (Vol. 16, No. 3/4, 1997). *"Presents a most satisfying collection of articles that will be of interest, first and foremost, to chemistry librarians, but also to science librarians working in other science disciplines within academic settings." (Barbara List, Director, Science and Engineering Libraries, Columbia University, New York, New York)*

History of Science and Technology: A Sampler of Centers and Collections of Distinction, edited by Cynthia Steinke, MS (Vol. 14, No. 4, 1995). *"A 'grand tour' of history of science and technology collections that is of great interest to scholars, students and librarians." (Jay K. Lucker, AB, MSLS, Director of Libraries, Massachusetts Institute of Technology; Lecturer in Science and Technology, Simmons College, Graduate School of Library and Information Science)*

Instruction for Information Access in Sci-Tech Libraries, edited by Cynthia Steinke, MS (Vol. 14, No. 2, 1994). *"A refreshing mix of user education programs and contain[s] many examples of good practice." (Library Review and Reference Reviews)*

Scientific and Clinical Literature for the Decade of the Brain, edited by Tony Stankus, MLS (Vol. 13, No. 3/4, 1993). *"This format combined with selected book and journal title lists is very convenient for life science, social science, or general reference librarians/bibliographers who wish to review the area or get up to speed quickly." (Ruth Lewis, MLS, Biology Librarian, Washington University, St. Louis, Missouri)*

Sci-Tech Libraries of the Future, edited by Cynthia Steinke, MS (Vol. 12, No. 4 and Vol. 13, No. 1, 1993). *"Very timely.... Will be of interest to all libraries confronted with changes in technology, information formats, and user expectations." (LA Record)*

Science Librarianship at America's Liberal Arts Colleges: Working Librarians Tell Their Stories, edited by Tony Stankus, MLS (Vol. 12, No. 3, 1992). *"For those teetering on the tightrope between the needs and desires of science faculty and liberal arts librarianship, this book brings a sense of balance." (Teresa R. Faust, MLS, Science Reference Librarian, Wake Forest University)*

Biographies of Scientists for Sci-Tech Libraries: Adding Faces to the Facts, edited by Tony Stankus, MLS (Vol. 11, No. 4, 1992). *"A guide to biographies of scientists from a wide variety of scientific fields, identifying titles that reveal the personality of the biographee as well as contributions to his/her field." (Sci Tech Book News)*

Information Seeking and Communicating Behavior of Scientists and Engineers, edited by Cynthia Steinke, MS (Vol. 11, No. 3, 1991). *"Unequivocally recommended.... The subject is one of importance to most university libraries, which are actively engaged in addressing user needs as a framework for library services." (New Library World)*

Technology Transfer: The Role of the Sci-Tech Librarian, edited by Cynthia Steinke, MS (Vol. 11, No. 2, 1991). *"Educates the reader about the role of information professionals in the multifaceted technology transfer process." (Journal of Chemical Information and Computer Sciences)*

Electronic Information Systems in Sci-Tech Libraries, edited by Cynthia Steinke, MS (Vol. 11, No. 1, 1990). *"Serves to illustrate the possibilities for effective networking from any library/information facility to any other geographical point." (Library Journal)*

The Role of Trade Literature in Sci-Tech Libraries, edited by Ellis Mount, DLS (Vol. 10, No. 4, 1990). *"A highly useful resource to identify and discuss the subject of manufacturers' catalogs and their historical as well as practical value to the profession of librarianship. Dr. Mount has made an outstanding contribution." (Academic Library Book Review)*

Role of Standards in Sci-Tech Libraries, edited by Ellis Mount, DLS (Vol. 10, No. 3, 1990). *Required reading for any librarian who has been asked to identify standards and specifications.*

Relation of Sci-Tech Information to Environmental Studies, edited by Ellis Mount, DLS (Vol. 10, No. 2, 1990). *"A timely and important book that illustrates the nature and use of sci-tech information in relation to the environment." (The Bulletin of Science, Technology & Society)*

End-User Training for Sci-Tech Databases, edited by Ellis Mount, DLS (Vol. 10, No. 1, 1990). *"This is a timely publication for those of us involved in conducting online searches in special libraries where our users have a detailed knowledge of their subject areas." (Australian Library Review)*

Sci-Tech Archives and Manuscript Collections, edited by Ellis Mount, DLS (Vol. 9, No. 4, 1989). *Gain valuable information on the ways in which sci-tech archival material is being handled and preserved in various institutions and organizations.*

Collection Management in Sci-Tech Libraries, edited by Ellis Mount, DLS (Vol. 9, No. 3, 1989). *"An interesting and timely survey of current issues in collection management as they pertain to science and technology libraries." (Barbara A. List, AMLS, Coordinator of Collection Development, Science & Technology Research Center, and Editor, New Technical Books, The Research Libraries, New York Public Library)*

The Role of Conference Literature in Sci-Tech Libraries, edited by Ellis Mount, DLS (Vol. 9, No. 2, 1989). *"The volume constitutes a valuable overview of the issues posed for librarians and users by one of the most frustrating and yet important sources of scientific and technical information." (Australian Library Review)*

Adaptation of Turnkey Computer Systems in Sci-Tech Libraries, edited by Ellis Mount, DLS (Vol. 9, No. 1, 1989). *"Interesting and useful. . . . The book addresses the problems and benefits associated with the installation of a turnkey or ready-made computer system in a scientific or technical library." (Information Retrieval & Library Automation)*

Sci-Tech Libraries Serving Zoological Gardens, edited by Ellis Mount, DLS (Vol. 8, No. 4, 1989). *"Reviews the history and development of six major zoological garden libraries in the U.S." (Australian Library Review)*

Libraries Serving Science-Oriented and Vocational High Schools, edited by Ellis Mount, DLS (Vol. 8, No. 3, 1989). *A wealth of information on the special collections of science-oriented and vocational high schools, with a look at their services, students, activities, and problems.*

Sci-Tech Library Networks Within Organizations, edited by Ellis Mount, DLS (Vol. 8, No. 2, 1988). *Offers thorough descriptions of sci-tech library networks in which their members have a common sponsorship or ownership.*

One Hundred Years of Sci-Tech Libraries: A Brief History, edited by Ellis Mount, DLS (Vol. 8, No. 1, 1988). *"Should be read by all those considering, or who are already involved in, information retrieval, whether in Sci-tech libraries or others." (Library Resources & Technical Services)*

Alternative Careers in Sci-Tech Information Service, edited by Ellis Mount, DLS (Vol. 7, No. 4, 1987). *Here is an eye-opening look at alternative careers for professionals with a sci-tech background, including librarians, scientists, and engineers.*

Preservation and Conservation of Sci-Tech Materials, edited by Ellis Mount, DLS (Vol. 7, No. 3, 1987). *"This cleverly coordinated work is essential reading for library school students and practicing librarians. . . . Recommended reading." (Science Books and Films)*

Sci-Tech Libraries Serving Societies and Institutions, edited by Ellis Mount, DLS (Vol. 7, No. 2, 1987). *"Of most interest to special librarians, providing them with some insight into sci-tech libraries and their activities as well as a means of identifying specialized services and collections which may be of use to them." (Sci-Tech Libraries)*

Innovations in Planning Facilities for Sci-Tech Libraries, edited by Ellis Mount, DLS (Vol. 7, No. 1, 1986). *"Will prove invaluable to any librarian establishing a new library or contemplating expansion." (Australasian College Libraries)*

Role of Computers in Sci-Tech Libraries, edited by Ellis Mount, DLS (Vol. 6, No. 4, 1986). *"A very readable text. . . . I am including a number of the articles in the student reading list." (C. Bull, Kingstec Community College, Kentville, Nova Scotia, Canada)*

Weeding of Collections in Sci-Tech Libraries, edited by Ellis Mount, DLS (Vol. 6, No. 3, 1986). *"A useful publication. . . . Should be in every science and technology library." (Rivernia Library Review)*

Sci-Tech Libraries in Museums and Aquariums, edited by Ellis Mount, DLS (Vol. 6, No. 1/2, 1985). *"Useful to libraries in museums and aquariums for its descriptive and practical information." (The Association for Information Management)*

Data Manipulation in Sci-Tech Libraries, edited by Ellis Mount, DLS (Vol. 5, No. 4, 1985). *"Papers in this volume present evidence of the growing sophistication in the manipulation of data by information personnel." (Sci-Tech Book News)*

Role of Maps in Sci-Tech Libraries, edited by Ellis Mount, DLS (Vol. 5, No. 3, 1985). *Learn all about the acquisition of maps and the special problems of their storage and preservation in this insightful book.*

Fee-Based Services in Sci-Tech Libraries, edited by Ellis Mount, DLS (Vol. 5, No. 2, 1985). *"Highly recommended. Any librarian will find something of interest in this volume." (Australasian College Libraries)*

Serving End-Users in Sci-Tech Libraries, edited by Ellis Mount, DLS (Vol. 5, No. 1, 1984). *"Welcome and indeed interesting reading. . . . a useful acquisition for anyone starting out in one or more of the areas covered." (Australasian College Libraries)*

Management of Sci-Tech Libraries, edited by Ellis Mount, DLS (Vol. 4, No. 3/4, 1984). *Become better equipped to tackle difficult staffing, budgeting, and personnel challenges with this essential volume on managing different types of sci-tech libraries.*

Collection Development in Sci-Tech Libraries, edited by Ellis Mount, DLS (Vol. 4, No. 2, 1984). *"Well-written by authors who work in the field they are discussing. Should be of value to librarians whose collections cover a wide range of scientific and technical fields." (Library Acquisitions: Practice and Theory)*

Role of Serials in Sci-Tech Libraries, edited by Ellis Mount, DLS (Vol. 4, No. 1, 1983). *"Some interesting nuggets to offer dedicated serials librarians and users of scientific journal literature. . . . Outlines the direction of some major changes already occurring in scientific journal publishing and serials management." (Serials Review)*

Planning Facilities for Sci-Tech Libraries, edited by Ellis Mount, DLS (Vol. 3, No. 4, 1983). *"Will be of interest to special librarians who are contemplating the building of new facilities or the renovating and adaptation of existing facilities in the near future. . . . A useful manual based on actual experiences." (Sci-Tech News)*

Monographs in Sci-Tech Libraries, edited by Ellis Mount, DLS (Vol. 3, No. 3, 1983). *This insightful book addresses the present contributions monographs are making in sci-tech libraries as well as their probable role in the future.*

Role of Translations in Sci-Tech Libraries, edited by Ellis Mount, DLS (Vol. 3, No. 2, 1983). *"Good required reading in a course on special libraries in library school. It would also be useful to any librarian who handles the ordering of translations." (Sci-Tech News)*

Online versus Manual Searching in Sci-Tech Libraries, edited by Ellis Mount, DLS (Vol. 3, No. 1, 1982). *An authoritative volume that examines the role that manual searches play in academic, public, corporate, and hospital libraries.*

Document Delivery for Sci-Tech Libraries, edited by Ellis Mount, DLS (Vol. 2, No. 4, 1982). *Touches on important aspects of document delivery and the place each aspect holds in the overall scheme of things.*

Cataloging and Indexing for Sci-Tech Libraries, edited by Ellis Mount, DLS (Vol. 2, No. 3, 1982). *Diverse and authoritative views on the problems of cataloging and indexing in sci-tech libraries.*

Role of Patents in Sci-Tech Libraries, edited by Ellis Mount, DLS (Vol. 2, No. 2, 1982). *A fascinating look at the nature of patents and the complicated, ever-changing set of indexes and computerized databases devoted to facilitating the identification and retrieval of patents.*

Current Awareness Services in Sci-Tech Libraries, edited by Ellis Mount, DLS (Vol. 2, No. 1, 1982). *An interesting and comprehensive look at the many forms of current awareness services that sci-tech libraries offer.*

Role of Technical Reports in Sci-Tech Libraries, edited by Ellis Mount, DLS (Vol. 1, No. 4, 1982). *Recommended reading not only for science and technology librarians, this unique volume is specifically devoted to the analysis of problems, innovative practices, and advances relating to the control and servicing of technical reports.*

Training of Sci-Tech Librarians and Library Users, edited by Ellis Mount, DLS (Vol. 1, No. 3, 1981). *Here is a crucial overview of the current and future issues in the training of science and engineering librarians as well as instruction for users of these libraries.*

Networking in Sci-Tech Libraries and Information Centers, edited by Ellis Mount, DLS (Vol. 1, No. 2, 1981). *Here is an entire volume devoted to the topic of cooperative projects and library networks among sci-tech libraries.*

Planning for Online Search Service in Sci-Tech Libraries, edited by Ellis Mount, DLS (Vol. 1, No. 1, 1981). *Covers the most important issue to consider when planning for online search services.*

Patent and Trademark Information: Uses and Perspectives

Virginia Baldwin
Editor

Patent and Trademark Information: Uses and Perspectives has been co-published simultaneously as *Science & Technology Libraries*, Volume 22, Numbers 1/2 2001.

The Haworth Information Press®
An Imprint of The Haworth Press, Inc.

Published by

The Haworth Information Press®, 10 Alice Street, Binghamton, NY 13904-1580 USA

The Haworth Information Press® is an imprint of The Haworth Press, Inc., 10 Alice Street, Binghamtom, NY 13904-1580 USA.

Patent and Trademark Information: Uses and Perspectives has been co-published simultaneously as *Science & Technology Libraries*™, Volume 22, Numbers 1/2 2001.

© 2001 by The Haworth Press, Inc. All rights reserved. No part of this work may be reproduced or utilized in any form or by any means, electronic or mechanical, including photocopying, microfilm and recording, or by any information storage and retrieval system, without permission in writing from the publisher. Printed in the United States of America.

The development, preparation, and publication of this work has been undertaken with great care. However, the publisher, employees, editors, and agents of The Haworth Press and all imprints of The Haworth Press, Inc., including The Haworth Medical Press® and Pharmaceutical Products Press®, are not responsible for any errors contained herein or for consequences that may ensue from use of materials or information contained in this work. Opinions expressed by the author(s) are not necessarily those of The Haworth Press, Inc. With regard to case studies, identities and circumstances of individuals discussed herein have been changed to protect confidentiality. Any resemblance to actual persons, living or dead, is entirely coincidental.

Cover design by Marylouise E. Doyle.

Front Cover: Photos of 19th century patent models obtained from the United States Patent and Trademark Office (USPTO), courtesy of the National Inventors Hall of Fame; facsimile image of the patent document obtained from the USPTO Web site, http://www.uspto.gov.

Library of Congress Cataloging-in-Publication Data

Patent and trademark information : uses and perspectives / Virginia Baldwin, editor.
    p. cm.
    Co-published simultaneously as Science & technology libraries.
    Includes bibliographical references and index.
    ISBN 0-7890-0425-9 (hard cover : alk. paper) – ISBN 0-7890-0440-2 (soft cover : alk. paper)
    1. Patents. 2. Patent searching. 3. Trademarks. 4. Trademark searching. I. Baldwin, Virginia A. II. Science & technology libraries.
T339.P238 2004
608–dc22

2003021257

Indexing, Abstracting & Website/Internet Coverage

This section provides you with a list of major indexing & abstracting services. That is to say, each service began covering this periodical during the year noted in the right column. Most Websites which are listed below have indicated that they will either post, disseminate, compile, archive, cite or alert their own Website users with research-based content from this work. (This list is as current as the copyright date of this publication.)

Abstracting, Website/Indexing Coverage	Year When Coverage Began
• *AGRICOLA Database (AGRICultural OnLine Access): A Bibliographic database of citations to the agricultural literature created by the National Agricultural Library and its cooperators* <http://www.natl.usda.gov/ag98>	1989
• *AGRIS*	1989
• *Aluminum Industry Abstracts* <http://www.csa.com>	2003
• *Biosciences Information Service of Biological Abstracts (BIOSIS) a centralized source of life science information* <http://www.biosis.org>	1982
• *BIOSIS Previews: online version of Biological Abstracts and Biological Abstracts/RRM (Reports, Reviews, Meetings); Covers approximately 6,500 life science journals and 2,000 worldwide meetings*	1982
• *Cambridge Scientific Abstracts is a leading publisher of scientific information in print journals, online databases, CD-ROM and via the Internet* <http://www.csa.com>	2003
• *Ceramic Abstracts* <http://www.csa.com>	2003
• *Chemical Abstracts Service–monitors, indexes & abstracts the world's chemical literature, updates this information daily, and makes it accessible through state-of-the-art information services* <http://www.cas.org>	1989
• *CNPIEC Reference Guide: Chinese National Directory of Foreign Periodicals*	1995
• *Computer and Information Systems Abstracts* <http://www.csa.com>	2003
• *Corrosion Abstracts* <http://www.csa.com>	2003
• *CSA Civil Engineering Abstracts* <http://www.csa.com>	2003
• *CSA Mechanical & Transportation Engineering Abstracts* <http://www.csa.com>	2003

(continued)

- Current Cites [Digital Libraries] [Electronic Publishing] [Multimedia & Hypermedia] [Networks & Networking] [General] <http://sunsite.berkeley.edu/CurrentCites/> 2000
- Current Index to Journals in Education 1995
- Educational Administration Abstracts (EAA) 1991
- Electronic and Communications Abstracts <http://www.csa.com> 2003
- Engineered Materials Abstracts (Cambridge Scientific Abstracts) <http://www.csa.com> 2003
- Engineering Information (PAGE ONE) 1987
- Environment Abstracts. Available in print–CD-ROM–on Magnetic Tape <http://www.cispubs.com> 1994
- FRANCIS. INIST/CNRS <http://www.inist.fr> 1993
- IBZ International Bibliography of Periodical Literature <http://www.saur.de> 1994
- Index Guide to College Journals (core list compiled by integrating 48 indexes frequently used to support undergraduate programs in small to medium sized libraries) 1999
- Index to Periodical Articles Related to Law 1990
- Information Science & Technology Abstracts: indexes journal articles from more than 450 publications as well as books, research reports, conference proceedings, and patents; EBSCO Publishing 1989
- Informed Librarian, The <http://www.infosourcespub.com> 1993
- INSPEC is the leading English-language bibliographic information service providing access to the world's scientific & technical literature in physics, electrical engineering, electronics, communications, control engineering, computers & computing, and information technology <http://www.iee.org.uk/publish/> 1982
- Journal of Academic Librarianship: Guide to Professional Literature, The 2000
- Konyvtari Figyelo (Library Review) 1995
- Library & Information Science Abstracts (LISA) <http://www.csa.com> 1989
- Library and Information Science Annual (LISCA) <http://www.lu.com> 1997
- Library Literature & Information Science <http://www.hwwilson.com> 1984
- Materials Business File–Steels Alerts <http://www.csa.com> 2003
- OCLC ArticleFirst <http://www.oclc.org/services/databases/> .... *
- OCLC ContentsFirst <http://www.oclc.org/services/databases/> .... *
- OCLC Public Affairs Information Service <http://www.pais.org> 1982
- PASCAL, c/o Institut de l'Information Scientifique et Technique. Cross-disciplinary electronic database covering the fields of science, technology & medicine. Also available on CD-ROM, and can generate customized retrospective searches <http://www.inist.fr> 1993
- Referativnyi Zhurnal (Abstracts Journal of the All-Russian Institute of Scientific and Technical Information–in Russian) 1982
- RESEARCH ALERT/ISI Alerting Services <http://www.isinet.com> 2000
- Solid State and Superconductivity Abstracts <http://www.csa.com> 2003
- Subject Index to Literature on Electronic Sources of Information <http://library.usask.ca/~dworacze/BIBLIO.HTM> 1996
- SwetsNet <http://www.swetsnet.com> 2001

*Exact start date to come.

(continued)

Special Bibliographic Notes related to special journal issues (separates) and indexing/abstracting:

- indexing/abstracting services in this list will also cover material in any "separate" that is co-published simultaneously with Haworth's special thematic journal issue or DocuSerial. Indexing/abstracting usually covers material at the article/chapter level.
- monographic co-editions are intended for either non-subscribers or libraries which intend to purchase a second copy for their circulating collections.
- monographic co-editions are reported to all jobbers/wholesalers/approval plans. The source journal is listed as the "series" to assist the prevention of duplicate purchasing in the same manner utilized for books-in-series.
- to facilitate user/access services all indexing/abstracting services are encouraged to utilize the co-indexing entry note indicated at the bottom of the first page of each article/chapter/contribution.
- this is intended to assist a library user of any reference tool (whether print, electronic, online, or CD-ROM) to locate the monographic version if the library has purchased this version but not a subscription to the source journal.
- individual articles/chapters in any Haworth publication are also available through the Haworth Document Delivery Service (HDDS).

Patent and Trademark Information: Uses and Perspectives

CONTENTS

Introduction 　*Virginia Baldwin*	1
PATENTS	
Patents for Victory: Disseminating Enemy Technical Information During World War II 　*Michael White*	5
The Seven Steps: Basic Novelty Patent Searching 　*Donna K. Hopkins*	23
Finding Grandpa's Patent: Using Patent Information for Historical or Genealogical Research 　*Jan Comfort*	39
esp@cenet®: Europe's Network of Patent Databases 　*Gerry McKiernan*	57
Regional Patent Systems: A Challenge for the International Searcher 　*Stephen R. Adams*	89
Patent Data for Technology Assessment, Part I: Applications, Patent Databases, and Retrieval Methods 　*Cynthia A. Kehoe* 　*Xiao Jason Yu*	101
Patent Data for Technology Assessment, Part II: Using U.S. Patent Data to Examine Trends in GPS Technology 　*Xiao Jason Yu* 　*Cynthia A. Kehoe*	117

TRADEMARKS

Finding Your Way Through the Trademark Information Maze   137
    Charlotte A. Erdmann

State Trademark and Company Name Web Sites   161
    James C. Miller

Index   175

ABOUT THE EDITOR

Virginia (Ginny) Baldwin, MS, MLS, is Associate Professor and Head of the Engineering Library and the Physics and Astronomy Library at the University of Nebraska in Lincoln. Ms. Baldwin is also the Patent and Trademark Librarian for the State of Nebraska. She is a former Scientific Programmer at Patrick Air Force Base in Florida, and an Engineer Specialist at Vandenberg Air Force Base in California. For nine years at Eastern Illinois University, she was responsible for collection development and specialized reference and library instruction in the engineering, computer science, and physical science disciplines. She was awarded academic tenure at Eastern Illinois University in 1997 and promoted to Professor in 1999.

Ms. Baldwin is the Editor of *Science & Technology Libraries*. She has been published in several journals, including *College and Research Libraries, Collection Management, Illinois Libraries, Journal of Technology Studies*, and *To Improve the Academy*, the Annual of the Professional and Organziational Development Network in Higher Education. She also authored a chapter in *Electronic Collection Management* (The Haworth Press, Inc.). She is the liaison from the Sci-Tech Division of Special Libraries Association to the Sci-Tech Section of the Association of College and Research Libraries, and is a member of both associations. Ms. Baldwin is also a member of the American Society for Engineering Education.

Introduction

There is a patent office at the seat of government of the universe, whose managers are as much interested in the dispersion of seeds as anybody at Washington can be, and their operations are infinitely more extensive and regular.

–Henry David Thoreau, "The Succession of Forest Trees" (1860), in *The Writings of Henry David Thoreau*, vol. 5, p. 187, Houghton Mifflin (1906)

Perchance, coming generations will not abide the dissolution of the globe, but, availing themselves of future inventions in aerial locomotion, and the navigation of space, the entire race may migrate from the earth, to settle some vacant and more western planet. . . . It took but little art, a simple application of natural laws, a canoe, a paddle, and a sail of matting, to people the isles of the Pacific, and a little more will people the shining isles of space.

–Henry David Thoreau. "Paradise (To Be) Regained" (1843), in *The Writings of Henry David Thoreau*, vol. 4, p. 292, Houghton Mifflin (1906)

PTDL's are empowered by the USPTO to manage knowledge of intellectual property information . . .

–Martha Crockett Sneed–former manager of the PTDL program at the USPTO

With the advent of United States patent and trademark information through the Internet, access is now readily available for librarians in academic libraries that are not depositories of this government information. Since each state has no more than one or a few Patent and Trademark Depository Libraries (PTDL), this United States Patent and Trademark Office (USPTO) achievement marks a new frontier in the Offices' goal of patent and trademark information access.

The information explosion has hit the intellectual property world. Now a researcher, inventor, genealogist, scientist, engineer, or any interested party can search United States, European, and international patent databases. All that is needed is an Internet connection and a Web browser. Patent literature has been defined as "grey literature" with the implication that it is difficult to locate a specific piece of the wealth of technological information that it contains.

This volume covers historical, practical, business, and research aspects of the use of patent and trademark information in the United States and in countries worldwide. While not covering the USPTO Web site (www.uspto.gov) in depth, articles on patent searching cover the basic steps. On trademark searching an article gives case examples and appropriate searches, and another gives an in-depth explanation of a European patent Web site. A patent searcher or librarian who reads this volume will be well-equipped to perform a preliminary patent search. Those readers who need further assistance with the USPTO web site can proceed, armed with the knowledge of the step by step process, to the help screen on the patent Web sites, to a Patent and Trademark Depository Librarian (locations are given in "Finding Grandpa's Patent: Using Patent Information for Historical or Genealogical Research") or to any librarian who is well versed in reading and interpreting database help screens.

An appropriate topic for the post "9-11" atmosphere still existent in 2003, this volume begins with an historical look at United States government seizure of intellectual property rights during World War II. In "Patents for Victory: Disseminating Enemy Technical Information During World War II," Michael White, an employee of the USPTO, traces the seizure of foreign-owned patents and the beginnings of the Patent and Trademark Depository Library Program as part of the Office of Alien Property Custodian's effort to accomplish the mission of making foreign patent information available to American industries for use in the war effort.

Before the creation of the USPTO Web site for patent and trademark searching, the USPTO outlined and distributed to Patent and Trademark

Depository Libraries a seven-step method to describe the patent searching process. In "The Seven Steps: Basic Novelty Patent Searching," Patent and Trademark Depository Program librarian Donna Hopkins describes the seven step process and translates that process into use of the databases that are currently available on the USPTO Web site. This article provides a basic understanding of intellectual property forms and clearly describes the key to patent "novelty" searching, that of determining the appropriate classes and subclasses for an invention. A good overview of the process, Hopkins provides details that even the experienced PTDL librarian may find useful.

The aforementioned article "Finding Grandpa's Patent: Using Patent Information for Historical or Genealogical Research" is an invaluable asset to the librarian who is asked to find information on historical patents. In this treatise Jan Comfort gives information on the intricacies of gleaning desired facts when often inadequate details are known by the patron. Comfort's article includes an annotated bibliography of print and electronic resources that provide a variety of potential clues for these types of searches and provides a list of PTDLs in which many of these resources reside.

In "*esp@cenet®*: Europe's Network of Patent Databases" Gerry McKiernan introduces us to the world of patents outside of the United States. He explains the structure of two classification systems at the European Patent Office and the language of the framework that has been established so that an inventor can more easily obtain protection in countries outside of his or her own. A must-read for those who want to extend their patent novelty search beyond the registrations of their own country, McKiernan's article includes many important database and search details obtained from communications with the European Patent Office. An important detail that is explained in his text is the greater range of bibliographic access to historical Unites States patents that is currently available through *esp@cenet®* as compared to the USPTO Web site.

Stephen Adams provides historical details about the origin and structure of regional patent offices and systems throughout the world in "Regional Patent Systems: A Challenge for the International Searcher." Adams differentiates between various regional systems, in terms of many aspects such as the application process itself, whether the regional system actually grants the patents, and whether the regional system will be involved in any litigation of its granted patents.

There are several databases that provide a variety of interfaces and complexity of search structures for United States patent data. In "Patent Data for Technology Assessment, Part I: Applications, Patent Databases, and Retrieval Methods" authors Cynthia A. Kehoe and Xiao Jason Yu describe database aspects within the framework of a variety of uses of patent information. In doing so, the stage is set for a follow-up description of an investigation of a specific area of technology, that of the global positioning system (GPS) in "Patent Data for Technology Assessment, Part II: Using U.S. Patent Data to Examine Trends in GPS Technology." In Part II the same authors show how patent data can be retrieved and mined to find patterns of development and innovations in an area of technology (GPS).

Two articles in this volume cover the (United States) world of trademark searching. In "Finding Your Way Through the Trademark Information Maze," Charlotte Erdmann accomplishes much. She gives basic guidance in performing a trademark search, compares two USPTO databases for trademark searching, and through a variety of sample scenarios, gives insight into a myriad of possibilities for finding value in trademark information. Erdmann's article covers national trademark databases. In "State Trademark and Company Name Web Sites" James Miller introduces us to the often overlooked necessity of performing state trademark and company name searches. Miller's research provides a complete snapshot in time of the search capabilities of state trademark agencies.

This volume emerges at a crucial moment in intellectual property information availability. The charge of the science and technology librarian has always been to proactively provide service. Now for the first time in history worldwide intellectual property information is available through an easily accessible interface, the World Wide Web. Our next step is an important one. Our researchers, faculty, and students must be kept informed of the importance of this tool and the importance of learning how to use it. Through individual patron contacts, library instruction, and through library print and web-based "handouts," the opportunity is now ours to spread the word that will provide to our company's scientists, our university researchers, and our students–the scientists and engineers of the future–knowledge of and access to an information resource that is critical in their pursuit of the means to engineer the future.

Virginia Baldwin

PATENTS

Patents for Victory: Disseminating Enemy Technical Information During World War II

Michael White

SUMMARY. In national emergencies governments have the power to suspend patent rights. During World War II, the U.S. government seized tens of thousands of foreign-owned patents and other forms of intellectual property. The agency responsible for this was the Office of the Alien Property Custodian (APC). The APC was charged with making the technical information disclosed in these patents available to American industries for use in the war effort. To accomplish this mission the agency's

---

Michael White is Librarian, Patent and Trademark Depository Library Program, U.S. Patent and Trademark Office, Washington, DC 20231 (E-mail: michael.white@uspto.gov).

The author would like to thank Mr. James C. Miller, Reference Librarian, University of Maryland, for his assistance in researching this article.

[Haworth co-indexing entry note]: "Patents for Victory: Disseminating Enemy Technical Information During World War II." White, Michael. Co-published simultaneously in *Science & Technology Libraries* (The Haworth Information Press, an imprint of The Haworth Press, Inc.) Vol. 22, No. 1/2, 2001, pp. 5-22; and: *Patent and Trademark Information: Uses and Perspectives* (ed: Virginia Baldwin) The Haworth Information Press, an imprint of The Haworth Press, Inc., 2001, pp. 5-22. Single or multiple copies of this article are available for a fee from The Haworth Document Delivery Service [1-800-HAWORTH, 9:00 a.m. - 5:00 p.m. (EST). E-mail address: docdelivery@haworthpress.com].

http://www.haworthpress.com/store/product.asp?sku=J122
© 2001 by The Haworth Press, Inc. All rights reserved.

Patent Information and Marketing Section established patent libraries nationwide; exhibited patent information at trade shows and conferences; published and sold catalogs, indexes and abstracts of seized patents; and distributed patent information to other agencies and libraries. *[Article copies available for a fee from The Haworth Document Delivery Service: 1-800-HAWORTH. E-mail address: <docdelivery@haworthpress.com> Website: <http://www.HaworthPress.com> © 2001 by The Haworth Press, Inc. All rights reserved.]*

KEYWORDS. Patents, patent applications, copyrights, trademarks, inventions, patent classification, Office of the Alien Property Custodian, APC, Department of Justice, patent office, World War I, World War II, Leo T. Crowley, James E. Markham, compulsory patent licensing, APC Patent Libraries, information dissemination programs, scientific and technical information

INTRODUCTION

Congress shall have the power to promote the progress of science and the useful arts, by securing for limited times to authors and inventors the exclusive right to their respective writings and discoveries. (*Constitution of the United States*, Article 1, Section 8, Clause 8)

At the height of last year's anthrax scare, the U.S. government discovered that its stockpile of medicines used against diseases such as anthrax and smallpox contained only a few million doses. Nervous public health officials admitted that this supply was inadequate to treat even one metropolitan area exposed to biological attack. The obvious solution was to mobilize the pharmaceutical industry and increase drug production as quickly as possible, but companies such as Bayer, the maker of the antibiotic CIPRO, balked at allowing competitors to make generic versions of their patented drugs. Many countries permit the compulsory licensing of pharmaceutical patents in public health emergencies, but the practice is unheard of in the U.S., which has traditionally supported strong patent protection for drug manufacturers (Chartrand 2001). The government of Canada acted decisively by suspending Bayer's patent rights and ordering one million doses from generic manufacturers. The U.S. response was uncertain and wavering. Health officials initially promised to respect Bayer's patents but then suggested that the government might resort to price controls or compulsory licensing in order to obtain the needed drugs (Bradsher 2001). Representative Sherrod Brown of Ohio went so far as to introduce legislation to that effect in November (Brown, H.R. 3235). Under this pressure and facing a public relations disaster, Bayer and other drug companies agreed to meet the government's price in exchange for keeping their patent rights intact.

This was not the first time that a national emergency has prompted the federal government to suspend or usurp the rights of inventors and patent owners. During World War II, the U.S. seized tens of thousands of patents and patent applications belonging to citizens of the Axis powers, Germany, Italy and Japan, their allies Romania, Hungary and Bulgaria, and Axis-occupied nations with the intent of making the technology disclosed in them available to U.S. industry. The agency authorized with this task was the Office of the Alien Property Custodian, commonly referred to as the APC. The APC had the power to seize or "vest" real property, such as ships, real estate, business enterprises, factories, personal property and all forms of intellectual property. From March 1942 to October 1946 the APC vested approximately 46,000 patents, 4,800 patent applications, 800 inventions, 400 trademarks and 200,000 copyrights, making it the largest holder of intellectual property in the U.S. at that time (APC Annual Report 1945, 1949). One of the most infamous assets seized was the copyright to Adolf Hitler's *Mein Kampf* (APC Vesting Order 128, 1942). Of course, the activities of the APC during World War II, especially with respect to the seizure of private property, are diverse and complicated. This paper will focus only on the agency's patent information dissemination and licensing programs. However, in order to understand these programs and their impact on the war effort it is necessary to first summarize the pre-war history of the APC.

WORLD WAR I AND THE CREATION OF THE APC

Seizing the rights to *Mein Kampf* may not have been the greatest literary coup of the 20th century but it must have pleased many Americans in the first difficult months of 1942. However, it would not have been possible without the assistance of another German leader, Kaiser Wihelm II, a generation earlier. Congress formally established the Office of the Alien Property Custodian in the Trading With the Enemy Act enacted in 1917 shortly after the United States declared war on Imperial

Germany and entered World War I as an ally of Great Britain, France and Italy. The mission of the APC was to seize enemy-owned property in the U.S. and dispose of it for the benefit of the war effort. Prior to the war, German financial and commercial interests in the U.S. were significant and widespread. The German chemical industry, nurtured by fifty years of government support and protectionist policies, enjoyed especially strong influence through its numerous U.S. subsidiaries, licensing agreements and patents. German artificial dyes brightened the clothes Americans wore and colored the food they ate. American doctors treated infant paralysis and other diseases with drugs created by the German companies such as Merck and Bayer. Other desirable German technology included optics, electrical equipment and scientific instruments, many of which were protected by U.S. patents. From 1900-1916, German inventors obtained more than 100,000 U.S. patents (Patent Office Annual Reports, 1900-1916). The coming of war in 1914 exposed the dependence of the U.S. economy on imported German technology and technical information.

Although the United States was neutral for the first three years of the war, the British maritime blockade of Germany effectively cut off American manufacturers and consumers from German technology and products. Germany, lacking the naval power to break the blockade, turned to unconventional means in order to maintain trade relations with the United States. On July 16, 1916, the German submarine *Deutschland* arrived in Baltimore with a million-dollar cargo of artificial dyes and pharmaceuticals. Financed by a group of German businessmen, the *Deutschland* was the first of seven large merchant submarines capable of carrying hundreds of tons of cargo. The sub's visit captivated the public, journalists and politicians. Captain Paul Koenig, the politically astute commander of the *Deutschland*, gave press interviews and entertained groups of U.S. senators and representatives while his crew loaded a cargo of American nickel, tin and crude rubber destined for German armament factories. The *Deutschland* made another trip to the U.S. in November 1916 carrying $10 million in drugs, dyes, securities and precious stones and returned to Germany with more raw materials and mail. Unfortunately for the burgeoning transatlantic undersea shipping business, the German government's decision in January 1917 to resume unrestricted submarine warfare, an act that enraged Americans, torpedoed any hope of continued peaceful relations. The U.S. declared war on Germany on April 6, 1917.

Although the APC confiscated more than $556 million in German-owned business enterprises, ships, stocks and intellectual property, its performance was hobbled by time, inexperience and scandal. Germany agreed to an armistice in November 1918, less than a year after the APC was organized. The government had taken no steps before the war to identify German-owned property and the APC had confiscated the property and goods of many naturalized German-American citizens by the time the guns ceased firing at 11:11 a.m. on November 11, 1918. The APC also undertook questionable sales and transfers of enemy property in its haste to dispose of its assets before a peace settlement was concluded. From 1917 to 1925 the APC sold 6,373 patents and licensed another 5,867 out of approximately 10,000 patents and patent applications seized (APC 1925). In early 1919, while President Wilson was in Europe attending the Versailles Peace Conference, Alien Property Custodian A. Mitchell Palmer and his successor Francis P. Garvan approved the sale of thousands of patents, patent applications and trademarks to the Chemical Foundation, Inc., a consortium of American chemical companies. It was later alleged that the Chemical Foundation was a front for German interests seeking to reclaim their patent portfolio. In fact, during the 1920s some German companies, such as Bosch, did create shadow subsidiaries and holding companies in an attempt to reestablish their technological dominance and to stifle U.S. innovation. The subsequent scandal saw Garvan, who also served on the board of directors of the Chemical Foundation, charged with fraud. Lawsuits and property claims, both legitimate and fraudulent, harried the APC into the 1930s (Steen 2001).

THE APC IN WORLD WAR II

In 1934, the APC became an agency of the Department of Justice where it continued to resolve claims and lawsuits arising from its activities during World War I. In December 1941, shortly after the Japanese attack on Pearl Harbor and Germany's declaration of war against the U.S., President Roosevelt asked Leo T. Crowley, a fellow Democrat and head of the Federal Deposit Insurance Corporation, to take on the job of Custodian. Roosevelt had been planning for more than a year to move the APC out of the Justice Department and establish it as an independent agency under his direct authority. He understood that the APC was a politically sensitive position and did not want a repeat of the scandals that had tarnished the agency and the administration of President Wilson (Weiss 1996). As late as 1943, the U.S. was still prosecuting claims made against property seized in 1918-1919 ("U.S. Sues Lawyer over Alien Assets" *The New York Times* 1943).

Roosevelt was also familiar with the problems of mobilizing war production when critical technologies were under patent protection. During World War I, the United States Army and Navy had experienced difficulties filling orders for equipment and spare parts because manufacturers were reluctant to accept government contracts for fear of infringing another company's patent rights. As acting Secretary of the Navy, Roosevelt had urged the Senate Naval Affairs Committee to take action to protect war contractors from patent infringement suits. Congress obliged by passing an amendment to the patent law that absolved contractors of infringement penalties and permitted patent holders instead to sue the government in the Court of Claims ("FDR vs. Arnold?" *Business Week* 1942).

Crowley agreed to lead the APC but had to wait for his appointment for almost four months while Roosevelt finessed his reorganization plan with Attorney General Francis Biddle and Secretary of the Treasury Henry Morgenthau, both of whom were eager to claim part of the alien property pie. Ultimately, Roosevelt succeeded and on March 11, 1942 he issued Executive Order 9095 that established the APC within the Office of Emergency Management. Several months later on July 6, 1942, Roosevelt issued another executive order that defined in detail the powers and duties of the APC in relation to the Department of the Treasury, which retained control over some enemy-owned financial assets.

Custodian Crowley quickly set his agency to work. The Division of Patent Administration, operating from offices in Washington, Chicago and New York, was responsible for all patent-related activities. The first task was to determine the number and current status of enemy-owned or controlled patents, patent applications and inventions. In May, the APC and Treasury Department issued orders forbidding enemy nationals from conducting patent-related business, for example filing applications, amendments, paying maintenance fees, etc. APC staff with the help of the Patent Office began compiling lists of foreign-owned patents and trademarks. In July, Crowley issued orders requiring any person claiming title to or interest in a patent or patent application granted to a foreign national to register their claims with the APC. These orders were published in the *Official Gazette of the U.S. Patent Office* on May 26 and July 14, respectively. Throughout the fall of 1942 the APC continued to identify and vest enemy patents and patent applications. By December, the APC had seized title to 30,000 enemy patents ("Free Patents" *Business Week* 1942). Ultimately, the APC vested over 50,000 patents and patent applications and exerted indirect control over hundreds more owned by vested corporations and patent holding companies. (See Table 1.)

On December 8, 1942, President Roosevelt announced that all foreign patents and applications under the control of the APC would be licensed to American industries. Custodian Crowley formally outlined the policy in an open letter to trade associations and industries released December 12, inspiring *Business Week* to dub him the "Patent Santa Claus" ("Patent Grab Bag" *Business Week* 1942). Under the policy the APC would issue royalty-free, non-exclusive licenses on all vested patents and applications. Each license application required an administration fee of $50 for the first patent and $5 for each additional patent covered by the license. In July 1943, the fee was reduced to a flat $15 per patent. Patents owned by citizens of enemy-occupied countries, France, Belgium, etc., would receive reasonable royalties beginning six months after the end of the war. The APC would collect and hold all royalties on behalf of pre-war patent owners until they could reclaim ownership. APC staff attorneys would continue prosecution on all vested patent applications in order to prevent them from becoming abandoned.

Not everyone agreed that nationalizing enemy patents was a just wartime expediency or in the long-term interest of the United States. While the practice of using captured enemy weapons, military supplies and installations had long been accepted among nations, the seizure of private property owned by citizens of enemy and enemy-occupied countries troubled some legal experts. Professor Edwin Borchard of Yale University warned that the seizure of enemy patents would undermine American property rights and investments abroad worth billions of dollars. "This country," wrote Borchard, "should therefore exert its influence to prevent the further corrosion of the institution of private property, since the United States and its citizens have more to lose by confiscation than any other country" (Borchard 1943).

The APC countered these arguments by pointing out the obvious advantages of removing any benefit to the enemy from the use of these patents and their strategic importance to the war effort. John Roe, assistant general council for the APC, wrote that "one of the most potent weapons we possess in our war against the Axis powers is our control over patents and patent applications which have been seized by the Alien Property Custodian, covering the most advanced developments in chemistry, electricity, meteorology, and other sciences and controlling processes that can be of tremendous value to America, both in the pros-

TABLE 1. Patents and Patent Applications Seized by the APC During World War II (APC 1952)

Residence of former owner	Total	Patents	Part Interests in Patents	Patent Applications	Abandoned Patent Applications	Inventions
Grand total	47,641	41,176	362	4,706	529	868
Enemy countries	34,662	29,905	282	3,418	435	622
Germany	30,855	26,713	246	2,962	379	555
Japan	1,215	1,126	1	72	14	2
Italy	1,920	1,561	16	259	29	55
Hungary	569	428	17	105	11	8
Rumania	73	59	2	8	2	2
Bulgaria	5	3	-	2	-	-
Two or more countries	25	15	-	10	-	-
Formerly enemy-occupied countries	12,934	11,231	79	1,285	93	246
Belgium	918	849	6	54	5	4
China (occupied)	3	-	-	3	-	-
Czechoslovakia	823	731	9	56	18	9
Denmark	586	516	2	59	-	9
Dutch East Indies	21	-	-	18	-	3
Estonia	11	11	-	-	-	-
France	8,133	7,125	41	731	48	188
Greece	13	10	-	1	2	-
Hong Kong	8	8	-	-	-	-
Latvia	11	11	-	-	-	-
Lithuania	4	4	-	-	-	-
Luxembourg	67	47	-	20	-	-
Monaco	1	-	-	1	-	-
Netherlands	1,455	1,156	3	256	15	25
Norway	674	607	8	50	2	7
Phillipine Islands	18	-	-	18	-	-
Poland	124	115	1	4	3	1
Staits Settlements	1	-	-	1	-	-
Thailand	1	1	-	-	-	-
Yugoslavia	30	26	1	3	-	-
Two or more countries	32	14	8	10	-	-

ecution of the war and in the economic development to follow" (Roe 1943). Mindful of the scandals over the sales of vested patents after World War I, Crowley declared that enemy patents would not be sold or transferred but held as "permanent positions of the American people" (MacCormac 1942).

Of course, German, Italian and Japanese patent owners, as declared enemies of the United States, had no means of challenging the government's compulsory licensing policy. But friendly Axis-occupied countries could and did voice their concerns. In May 1943, Norway, Belgium and the Netherlands lodged formal protests with the State Department over the issuing of royalty-free licenses on patents owned by their citizens ("Axis-Held Nations Hit Patents Policy" *The New York Times* 1943). At issue was not the licensing of patents required for wartime production, but the fact that all vested patents were eligible whether or not they disclosed strategically important technology. The APC's policy applied to patents on automatic towel dispensers and automatic gun loaders alike. Another point of concern was the fact that the licensed patents would generate no royalties for their owners until six months after the end of the war. All three governments urged the APC to collect royalties for the duration of the license. Eventually, the APC agreed with this argument and began collecting royalties on the appropriate patents.

Despite these concerns, intellectuals and the press generally supported the use of enemy patents and, in some cases, even the compulsory licensing of U.S. patents. In the spring and summer of 1942, journalist I. F. Stone wrote scathing columns in *The Nation*, a prominent leftist journal, accusing U.S. companies of zealously protecting their patent interests at the expense of the war effort. Specifically, Stone condemned Standard Oil for its reluctance to share synthetic fuel patents and trade secrets with the Soviet Union when thousands of Soviet soldiers were sacrificing their lives for the Allied cause ("Russian Lives and Oil Patents" Stone 1942). *The New Republic*, another leading weekly, argued for compulsory licensing of all patents, U.S. and foreign ("Senate Committee Blazes a Trail" *The New Republic* 1942). Some radical proposals reached far beyond wartime needs and recommended that the U.S. should adopt the Soviet practice of paying inventors a stipend for their discoveries in lieu of granting them patent rights. *The New Republic* belittled the reform proposals of Roosevelt's National Patent Planning Commission, established in December 1941 and chaired by Dr. Charles Kettering, president of General Motors and a prolific inventor ("Whitewashing the Patent System" *New Republic* 1943).

These views seems extreme now, but in the 1930s and 1940s many Americans, members of Congress and government officials believed that patents were a dangerous form of monopoly that benefited only large companies, discouraged innovation and stifled free trade. The Temporary National Economic Committee (TNEC) of the late 1930s proposed changes in U.S. patent law that would have effectively ended privately held patent rights altogether. In the Department of Justice, Assistant Attorney General Thurman W. Arnold was well known as a patent "buster" for his unrelenting pursuit of alleged abusive patent monopolies and foreign patent cartels. During 1942, Thurman repeatedly blamed patent monopolies for creating bottlenecks in war production (Thurman 1942). Subsequent government investigations did not uphold his allegations. In Congress, supporters of compulsory licensing included senators Joseph O'Mahoney of Wyoming, former chairman of the TNEC, Homer Bone of Washington and Robert La Follette of Wisconsin, and representative Jerry Voorhis of California ("For Free Patents" *Business Week* 1942). The debate over patent monopolies and inventor rights continued in the press, professional journals and Congress through the end of the war.

THE APC INFORMATION DISSEMINATION PROGRAM

The success of the APC's patent licensing program depended on a well-organized and active information dissemination campaign targeting potential licensees, namely manufacturers. The initiative, spearheaded by the Patent Marketing and Information Section assigned to the APC field office in Chicago, was well underway even before Crowley's formal announcement in December 1942. During the fall of 1942, Section staff cataloged, abstracted and classified vested patents and applications. A corresponding index arranged by U.S. Patent Classification was also compiled. APC staff first distributed lists of vested patents at the National Chemical Exhibition in Chicago in November 1942. According to one report, the exhibit attracted many technicians and patent attorneys who ordered copies of patent catalogs ("Free Patents" *Business Week* 1942). The two-volume catalog could be purchased for $5.00; individual sections covering one patent classification cost 10 cents each and 25 cents each for the nine largest classifications. A complete set of abstracts of mechanical and electrical patents in five volumes sold for $25. By June 1944 the APC had sold 3,110 copies of the full catalog and 5,882 copies of individual sections (APC 1944).

Professional associations contributed their expertise to help the APC organize and disseminate enemy patent information. During 1944, the Chicago Section of the American Chemical Society and the Science-Technology Group of the Special Libraries Association compiled a thirty-four volume set of abstracts plus an index covering more than 6,000 chemical patents. Copies were sold for $25. In January 1945, the APC published a supplement covering an additional 1,700 chemical patents.

In January 1943, the APC authorized the Patent Office to publish vested patent applications. The Patent Office published lists of vested applications arranged by classification and serial number in the *Official Gazette*, the office's weekly periodical announcing new patents. Applications in Class 260, organic chemistry, were published in the *Official Gazette* on April 20, 1943. Classes 1 to 64 on April 27, and Classes 4 to 102 on May 4. Additional applications in all classes were published on May 11, 18 and 25, June 26 and July 13. The public could order from the Commissioner of Patents copies of published applications for 10 cents each.

In order to allow public inspection of vested patents and patent applications, the APC established patent libraries in its field offices in Chicago, New York and Washington, D.C. Each library was staffed by "competent" librarians and stocked with a full set of catalogs, abstracts, indexes, Patent Office classification manuals and other searching aids. APC librarians were trained to help the public conduct searches and to prepare license applications. In 1944, the APC added additional libraries in Boston and Portland. In 1945, the library network expanded again to include Los Angeles, Kansas City and San Francisco. In addition to static libraries, the APC organized a traveling library that appeared at trade shows, professional meetings, chambers of commerce and factories.

In addition to the Patent Office, the APC cooperated with other government agencies and local organizations to disseminate patent information. In December 1943, the Bonneville Power Administration (BPA) convinced the APC to establish a library in the offices of its Marketing Division in Portland, Oregon. BPA was established as a federal agency in 1937 to manage hydroelectric projects and electricity supplies in the Pacific Northwest. The BPA marketing staff launched an aggressive promotional campaign that included luncheons in Seattle, Spokane and Portland to acquaint local businesses with the information contained in vested patents and a factory-to-factory tour through Washington, Oregon and Idaho.

In July 1943, the APC enlisted the Smaller War Plants Corporation (SWPC), the predecessor to the Small Business Administration, to disseminate enemy patent information. Approximately 100 SWPC district offices received copies of the complete catalog of vested patents and abstracts of non-chemical patents. The SWPC also offered technical assistance and financing to patent licensees (*The New York Times* 1943). The APC distributed patent catalogs to hundreds of chambers of commerce across the U.S.

The APC and Patent Office distributed copies of vested patents, catalogs and abstracts, but not published applications, to libraries that were designated as Patent Depository Libraries. The Patent Office first began distributing copies of patents to these libraries in 1871. Public libraries that received vested patents, catalogs and abstracts included:

Los Angeles, California
Chicago, Illinois
Boston, Massachusetts
Detroit, Michigan
St. Louis, Missouri
Newark, New Jersey
Albany, New York (State Dept. of Education)
Buffalo, New York (Grosvenor Library)
New York, New York
Cincinnati, Ohio
Cleveland, Ohio
Columbus, Ohio
Toledo, Ohio
Philadelphia, Pennsylvania (Franklin Institute)
Pittsburgh, Pennsylvania (Carnegie Library)
Providence, Rhode Island
Madison, Wisconsin (State Historical Society)

IMPACT OF THE APC PATENT INFORMATION AND LICENSING PROGRAMS

In mid 1944, political reasons forced President Roosevelt to accept the resignation of Leo T. Crowley as APC. His replacement was James E. Markham who served until October 1946 when President Truman abolished the APC as an independent agency and returned its employees and assets to the Department of Justice where it was known as the Alien Property Office. However, the defeat of the Axis in 1945 and Truman's reorganization order of 1946 did not end the agency's patent-related activities. The office continued to prosecute vested patent applications, license enemy patents, and distribute patent information well into the 1950s. Of the 4,802 patent applications vested during the war, 3,200 were issued as patents. Approximately 500 applications were abandoned (APC 1944). In the post-war period, the agency was also responsible for returning millions in patent royalties owed to citizens of friendly enemy-occupied countries and disposing of claims on vested patents, trademarks and copyrights.

The contribution of the agency's patent information dissemination and licensing programs to the U.S. war effort is difficult to measure. The net value of enemy property vested, excluding intellectual property, totaled $232 million (APC 1949). Custodian Markham reported to the president in 1945 that his agency had issued licenses covering 5,853 patents and patent applications as of June 30, 1944 (JPOS 1945). The majority of these licenses were issued on patents issued to German inventors. Noteworthy licensed inventions included Atabrine, a patented drug used to treat malaria, tungsten carbide, high-strength rayon, optical heat sensors, a process for producing wood alcohol from wood scraps and an automatic anti-aircraft gun loading device. From 1946 to 1952, the APO issued additional licenses on thousands of enemy patents and a handful of patents owned by citizens of enemy-occupied nations (APC 1949-1952).

Custodian Markham curtailed the agency's patent information activities in the months following the end of the war. The APC began dismantling its field offices and patent libraries in the spring of 1946. By June 1946, APC field offices in Chicago, Boston, Kansas City and Los Angeles had closed and their patent files had been discarded, archived or transferred to a local public library. At least two public libraries, Boston and Los Angeles, received parts of APC patent library collections (JPOS 1946). The Department of Justice maintained patent libraries at APO offices in New York, San Francisco and Washington through 1952.

The agency continued to sell at reduced prices catalogs and abstracts of vested patents. Attorney General Tom C. Clark promoted their value as technical references for libraries of research organizations, universities, inventors, law firms, and chambers of commerce (JPOS 1946). From 1949 to 1952, the APO received over 9,000 requests for information on vested patents and filled 3,025 orders for sets and sections of patent abstracts (APC 1949-1952). In addition to disseminating patent information, the agency distributed thousands of foreign scientific and technical books and periodicals through its copyright licensing and republication program. One of the best-known works republished was Beilstein's *Handbook on Organic Chemistry*, which before the war cost

$1,800 to $2,000. APC-licensed copies sold for $400. Other APC republished works included Gmelin's *Handbook on Inorganic Chemistry* and Wien-Hams' *Handbook of Experimental Physics*. The APC also issued copyright licenses for a small number of plays, motion pictures and music (APC 1943, 1944, 1952).

APC DOCUMENTS TODAY

The wartime records of the APC are located in the National Archives and Records Administration facility in College Park, Maryland. According to national online catalogs, a few libraries still hold copies of APC patent catalogs, indexes and abstracts. Vested applications that were issued as patents became permanent additions to the patent search files maintained by the U.S. Patent and Trademark Office. For example, U.S. patent 2,368,939, issued Feb. 6, 1945; filed Jan. 22, 1936, U.S. serial number 60,192; and published July 13, 1943. Vested applications that became abandoned are more difficult to locate. To the best of the author's knowledge, they were never converted to electronic format. The only known collection of APC patent applications open to the public is located in the Patent Search Room of the U.S. Patent and Trademark Office in Arlington, Virginia. The set is bound and arranged numerically by serial number. Published APC applications appear very similar in format to regular patents but are identified by their serial numbers and dates of publication. For example, "Published May 25, 1943 by the A.P.C." The assignee usually is listed as the "Alien Property Custodian" for patents that issued from 1944 to 1946 or the "Attorney General of the United States" for patents issued after 1946. According to the *Manual of Patent Examining Procedure* (MPEP), Sec. 901.06(c), APC applications are cited as follows: A.P.C. Application of ...., Ser. No............, Published .... It is unknown whether any additional collections of published APC applications exist in other libraries.

While the role and activities of the Alien Property Custodian relating to enemy-owned patents during World War II may be controversial, the APC's patent information dissemination program was successful in distributing large amounts of technical information to U.S. industry and the general public. Although not the first or the largest information dissemination program undertaken by a government agency, the APC's program demonstrated the importance and value of accessible patent information. Our wartime ally the Soviet Union understood this point very well. On one day in May 1946, Soviet representatives visited the Public Search Room of the U.S. Patent Office and ordered copies of 275,000 recently declassified patents ("Patents for Moscow" *Business Week* 1946). Nearly sixty years later, and facing a very different crisis, we are still reminded of this fact.

BIBLIOGRAPHY

Articles

"Alien Assets Frozen Total 7 Billions." *The New York Times*, May 6, 1943, p. 11.
"Alien-Owned Patents." *Science–Supplement*, vol. 96, no. 2503, December 18, 1942, p. 10.
"Alien Patents Listed." *The New York Times*, May 1, 1943, p. 25.
"Alien Properties Helped Win War." *The New York Times*, May 22, 1945, p. 28.
"Alien Property Custodian." *Journal of the Patent Office Society*, vol. 25, no. 1, January 1943, p. 57-71.
"Alien Property Custodian." *Journal of the Patent Office Society*, vol. 25, no. 6, June 1943, p. 433-434.
"Alien Property Custodian." *Journal of the Patent Office Society*, vol. 26, no. 12, December 1944, p. 819.
"Alien Property Custodian." *Journal of the Patent Office Society*, vol. 27, no. 4, April 1945, p. 283.
"Alien Property Custodian." *Journal of the Patent Office Society*, vol. 27, no. 6, June 1945, p. 403.
"Alien Property Custodian." *Journal of the Patent Office Society*, vol. 28, no. 6, June 1946, p. 426.
"Alien Property Custodian." *Journal of the Patent Office Society*, vol. 28, no. 9, September 1946, p. 686-687.
"Alien Property Custodian Establishes Patent Libraries." *The New York Times*, March 24, 1943, p. 29.
"Alien Property Office Ended by President." *The New York Times*, October 15, 1946, p. 13.
Anderson, John W. "Our Patent System Protects the Small Manufacturers." *Vital Speeches of the Day*, vol. 9, no. 12, April 1, 1943, p. 365-270.
Andrews, Edmund L. "Drug Maker Seems Uncertain in Response to Cipro Frenzy." *The New York Times*, October 20, 2001, Section A, p. 1.
Arnold, Thurman W. "The Abuse of Patents." *The Atlantic*, vol. 170, no. 1, July 1942, p. 14-20.
Arnold, Thurman W. "We Must Reform Patent Law." *The Atlantic*, vol. 170, no. 5, September 1942, p. 47-55.
"Atomic Energy Data." *Journal of the Patent Office Society*, vol. 27, no. 12, December 1945, p. 862-863.
"Axis-Held Nations Hit Patents Policy." *The New York Times*, May 13, 1943, p. 4.

Bradsher, Keith and Andrews, Edmund L. "U.S. Says Bayer Will Cut Cost of Its Anthrax Drug." *The New York Times*, October 24, 2001, Section B, p. 7.

Borchard, Edwin. "The Nationalization of Enemy Patents." *The American Journal of International Law* 37, (1943): 92-97.

Chartrand, Sandra. "Patents; In Health Emergencies, Brazil Allows the Copying of Drugs, to the Dismay of American Companies." *The New York Times*, February 19, 2001.

"Complete File of APC Patents Available in Los Angeles Public Library." *The Journal of the Patent Office Society*, vol. 28, no. 9, September 1946, p. 622.

"Crew Hears Prayer for U-Boat's Safety." *The New York Times*, July 17, 1916, p. 8.

"Department of Justice: Enemy Patent Abstracts." *Journal of the Patent Office Society*, vol. 28, no. 11, November 1946, p. 844-845.

"The *Deutschland* Eluded Foe With $10,000,000 Cargo." *The New York Times*, November 2, 1916, p. 1, 4.

"Enemy Patents: U.S. Publishes List of 8,000 Seized From Germans." *The New York Times*, January 18, 1945, p. B9.

"FDR vs. Arnold?" *Business Week*, May 23, 1942, p. 81.

"Fee on Foreign Patents Reduced." *The New York Times*, July 10, 1943, p. 16.

"For Free Patents." *Business Week*, April 11, 1942, p. 14.

"Free Patents." *Business Week*, December 19, 1942, p. 19-20.

Grahame, Arthur. "No More Fortunes for Inventors?" *Popular Science*, vol. 143, no. 6, December 1943, p. 78-81.

Hackley, Roy C. "Invention is Vital." *The Atlantic*, vol. 170, no. 4, October 1942, p. 49-55.

Haight, George I. "Patents and the General Welfare: the Right of an Inventor to His Invention." *Vital Speeches of the Day*, vol. 11, no. 10, March 1, 1945, p. 317-320.

Langer, Lawrence. "We Depend on Invention: an Answer to Thurman Arnold." The Atlantic, vol. 170, no. 1, July 1942, p. 21-30.

Leviero, Anthony. "Alien Goods Office to Lose Autonomy." *The New York Times*, September 25, 1946, p. 13.

MacCormac, John. "Enemy Patents Go to Our Industries." *The New York Times*, December 19, 1942, p. 20.

Montague, Gilbert H. "Foreign Trade and Patent Agreements." *Vital Speeches of the Day*, vol. 10, no. 17, June 1, 1944, p. 511-512.

"Organizing a War Economy." *The New Republic*, vol. 106, no. 18, May 4, 1942, p. 591.

"Patent Grab Bag." *Business Week*, December 21, 1942, p. 62.

"Patent Plum." *Business Week*, December 12, 1942, p. 8.

"Patent Reform." *Business Week*, June 26, 1943, p. 32-33.

"Patents and Free Enterprise." *Scientific American*, May 1942, p. 238-239.

"Patents Attacked." *Business Week*, November 27, 1943, p. 17-18.

"Patents for Moscow." *Business Week*, May 6, 1946, p. 30.

"Patents Go West." *Business Week*, January 8, 1944, p. 60-61.

"Patents on the March." *Business Week*, January 24, 1942, p. 52.

"Pending on Patents." *Business Week*, April 25, 1942, p. 20-21.

Pollack, Andrew. "Drug Makers Wrestle with World's New Rules; A Delicate Balance: Patriotism vs. Business." *The New York Times*, October 21, 2001.

Roe, John Ernest. "War Measures, The Alien Property Custodian and Patents." *Journal of the Patent Office Society* 25, (1943): 692-758.

"Secrecy versus Patents." Scientific American, March 1943, p. 113.

Steen, Kathryn. "Patents, Patriotism, and 'Skilled in the Art': USA v. The Chemical Foundation, Inc., 1923-1926." Isis, vol. 92, no. 1, March 2001, p. 91-122.

Stone, I. F. "Our Slacker Patents." *The Nation*, vol. 154, no. 18, May 2, 1942, p. 506.

Stone, I. F. "Russian Lives and Oil Patents." *The Nation*, vol. 155, no. 13, September 26, 1942, p. 261.

Stone, I. F. "The Truth About Rubber." *The Nation*, vol. 154, no. 16, April 18, 1942, p. 451.

"Senate Committees Blaze a Trail." *The New Republic*, vol. 106, no. 17, April 27, 1942, p. 572.

"Tug-of-War Fails to Halt Patent Reform Movement." *Business Week*, September 15, 1945, p. 7.

"U.S. Sues Lawyer over Alien Assets." *The New York Times*, January 9, 1943, p. 8.

"U.S. Takes Over from the Enemy." *Business Week*, June 6, 1942, p. 15.

Voorhis, Jerry (Representative from California). "The Patent Grant: Remedies to Prevent Its Monopolistic Abuse." *Vital Speeches of the Day*, vol. 11, no. 10, March 1, 1945, p. 315-317.

"War and Peace and the Patent System." *Fortune*, no. 26, August 1942, p. 103-105.

"Whitewashing the Patent System." *The New Republic*, vol. 109, no. 9, August 30, 1943, p. 278-279.

"Will Broaden Use of Enemy Patents." *The New York Times*, July 30, 1943, p. 21.

Books and Documents

The Alien Property Custodian: A Legislative History Chronological History and Bibliography of the Trading with the Enemy Act, 50 U.S. Code App. 1-40, and the Operations of the Office of Alien Property Custodian, 1917-1952. U.S. Senate, Committee on the Judiciary (82nd Congress, 2d Session).

Alien Property Custodian. "Vesting Order 128: All rights of Adolf Hitler, Franz Eher Nachf. GmbH, and Ferdinand Hirt in the copyrights covering *Mein Kampf*." 1942.

Alien Property Custodian. *Annual Report of the Office of Alien Property Custodian*. Washington, D.C.: Office of the Alien Property Custodian, 1921, 1923, 1925, 1944 and 1945.

American Chemical Society, Chicago Section. *Abstracts of Chemical Patents Vested in the Alien Property Custodian*. Washington: Office of the Alien Property Custodian, 1944. (Classified and indexed by the Science-Technology Group of the Special Libraries Association.)

Attorney General of the United States. *Annual Report of the Office of Alien Property*. Washington, D.C.: Department of Justice, 1949, 1950, 1951 and 1952.

Commissioner of Patents. *Annual Report*. Washington, D.C.: U.S. Patent Office, 1900-1918.

Crowley, Leo T. "General Order No. 2." *The Official Gazette of the United States Patent and Trademark Office*, vol. 540, no. 2, July 14, 1942, p. 235.

Crowley, Leo T. "General Order No. 3." *The Official Gazette of the United States Patent and Trademark Office*, vol. 540, no. 2, July 14, 1942, p. 235.

Division of Patent Administration. *Catalog of Vested Patents*. Washington: Office of the Alien Property Custodian, 1943.

Division of Patent Administration. *Index and Guide to Enemy Patents Vested in the Alien Property Custodian as of January 1946*. Washington: Office of the Alien Property Custodian, 1946.

Domke, Martin. *Trading With the Enemy in World War II*. New York: Central Book Co., 1943.

Domke, Martin. *The Control of Alien Property: Supplement to Trading with the Enemy in World War II*. New York: Central Book Co., 1947.

Henry, Conder C. "Notices." *The Official Gazette of the United States Patent Office*, vol. 538, no. 4, May 25, 1942, p. 739, 748.

Weiss, Stuart L. *The President's Man: Leo Crowley and Franklin Roosevelt in Peace and War*. Carbondale, Il: Southern Illinois University, 1996.

The Seven Steps:
Basic Novelty Patent Searching

Donna K. Hopkins

SUMMARY. A patent novelty search (to determine if an invention is new) consists of seven basic steps that can be summarized as determining the classification(s), finding patent numbers in those classifications, and then comparing the invention to the search results. A brief comparative review of the various types of intellectual property, patent basics, and the author's viewpoint regarding using keywords for this type of search are also included. *[Article copies available for a fee from The Haworth Document Delivery Service: 1-800-HAWORTH. E-mail address: <docdelivery@haworthpress.com> Website: <http://www.HaworthPress.com> © 2001 by The Haworth Press, Inc. All rights reserved.]*

KEYWORDS. Patent searching, patent databases, seven steps

The United States Patent and Trademark Office (USPTO) recommends a seven-step procedure for inventors and others who wish to search all of the U.S. patent literature. For the computer-phobic, all but one of the steps are possible (at some PTDLs) in an alternate media: print, microfiche, or microfilm. The exception to this is that since 2001 U.S. patents are only available in electronic form (both on the Web and on DVD-ROM).

Donna K. Hopkins is Engineering Librarian, Folsom Library, Rensselaer Polytechnic Institute (E-mail: hopkind@rpi.edu).

[Haworth co-indexing entry note]: "The Seven Steps: Basic Novelty Patent Searching." Hopkins, Donna K. Co-published simultaneously in *Science & Technology Libraries* (The Haworth Information Press, an imprint of The Haworth Press, Inc.) Vol. 22, No. 1/2, 2001, pp. 23-38; and: *Patent and Trademark Information: Uses and Perspectives* (ed: Virginia Baldwin) The Haworth Information Press, an imprint of The Haworth Press, Inc., 2001, pp. 23-38. Single or multiple copies of this article are available for a fee from The Haworth Document Delivery Service [1-800-HAWORTH, 9:00 a.m. - 5:00 p.m. (EST). E-mail address: docdelivery@haworthpress.com].

http://www.haworthpress.com/store/product.asp?sku=J122
© 2001 by The Haworth Press, Inc. All rights reserved.

All of the research tools are available at Patent and Trademark Depository Libraries (PTDLs) supported by the USPTO, and trained staff are available to help with their use. These tools, including the CASSIS disks,[1] are also available at some Government Printing Office (GPO) Depository Libraries. A list of item numbers is provided in the Appendix for selective depositories who want to select patent and trademark resources.

INTELLECTUAL PROPERTY BASICS

Before describing the specific steps, it may be useful to review some of the basics of intellectual property (IP) and of patents specifically. There are four basic types of intellectual property protection: trade secret, copyright, trademark, and patent. In the United States, the basis for two of these is covered in the Constitution, Article 1, Section 8, which gives Congress the power ". . . to promote the progress of science and useful arts, by securing for limited times to authors and inventors the exclusive right to their respective writings and discoveries." Trademarks and trade secrets were first covered by legislation in the late nineteenth century. Except for trade secrets, each has multiple types, and several kinds of protection (even multiple types of each) may be appropriate for a single "idea." At one time the game Monopoly® was protected by trademark, copyright, a utility patent, and a design patent.

Trade secrets are just that, secret. The protection lies in the fact that no one else knows. Companies can (and do) require confidentiality from employees, and take legal action if agreements are broken. The formula for Coca-Cola® could have been patented, but by now it would be public domain, which would make it legal for anyone to manufacture and sell the exact same recipe.

Copyright is the form of IP usually familiar to most librarians. In simplest terms it is the protection of an original *expression* of an idea (rather than the idea itself). This protection can cover as diverse forms as a sculpture or a Web page; a treatise on the political aspects of the economy or a motion picture; a musical recording or a computer program; a photograph or the choreography of a ballet. More specifics on the various types can be found at the Web site of the Library of Congress,[2] which administers copyright in the United States.

Patents and trademarks are both administered by the USPTO, which can be a source of confusion. The best way to keep them straight is to remember that trademarks function to identify the source or origin of products or services. This is done primarily with brand names and logos. Federal registration (that allows the use of the ®) is not required to use a trademark, and in fact may be refused for several reasons, such as generic terms or marks already in use.

Patents are granted on "inventions" and protect the actual technology involved. Inventors are given "exclusive" rights to what is covered in the patent. This means he or she can prevent others from making, using, selling, offering for sale, or importing the invention. A patent does not give the patent owner the right to make (or sell, etc.) the invention. For example, drugs will still need FDA approval.

SOME PATENT BASICS

There are several different types of patents granted in the United States. Most other countries have comparable types–some even have more. For those who want more detail on this, the British library has published an excellent book that describes all the "industrial property" publications from various countries.[3]

There are several types of U.S. patents that may appear in search results. With the exception of utility patents, which have no letter designation, each type of patent is designated by one or two letters (see Figure 1). The designating letter(s) appears at the beginning of the patent number; for example, D 405,111 is a design patent. Utility patent numbers do not begin with a letter. Any of these kinds of patents may also be prefaced by US to indicate that it is a United States patent, especially in databases that include non-US patents. For example US D405, 111 is a design patent granted by the USPTO, and US 6,000,000 is a utility patent granted by the USPTO.

What can be patented? For the most part, this concerns utility patents. There are two ways to answer this question. First, the invention must be

FIGURE 1. Types of Patent Documents

Plant patents (PP)	granted to asexually reproduced biological plants
Design patents (D)	protect appearance only (decorative)
Utility patents	"normal" patents, discussed in more detail in the rest of this article
Defensive Publications (T)	patents which were immediately in the public domain
Statutory Invention Registrations (H)	the current type of defensive publication

patentable subject matter. This is defined in the *United States Code* as "any new and useful process, machine, manufacture, or composition of matter, or any new and useful improvement thereof."[4] Second, it must meet three statutory requirements–it must be new, useful, and non-obvious. The "new" (or novelty) requirement is the reason most inventors want to search the patent databases and is defined as:

- must not have been described in a printed publication or patented anywhere, or been in public use or on sale in this country before the date the applicant made his invention

AND

- must not have been described in a printed publication anywhere, or been in public use or on sale in this country more than one year prior to the application for a US patent.[5]

A true patentability search includes an opinion (whether stated or merely implied) of whether or not a patent will result if an application is filed. It also involves any and all related literature (journals, product catalogs, etc.) and so involves the participation of a patent professional, preferably with knowledge of the industry involved. All previous literature (patents, articles, papers, etc.) that "read on"[6] even part of the invention is considered "prior art"–thus patentability searches can also be called prior art searches.

However, before reaching that stage, an inventor can perform a preliminary search by following the seven steps recommended by the USPTO. The steps are based on the resources used, and can be grouped into three basic concepts: determine the classification, find the patents in the classification, and then compare the resulting patents to the invention. Classification is also useful in other research where the subject matter of the patents is important, such as investigating the history of a technology or in competitive intelligence.

CLASSIFICATION

Step One: Index of U.S. Patent Classification

In preparation for using the *Index*, make a list of terms describing the invention. What is it? What does it do? What is it made of? How is it made? What makes it new and different? Also list some synonyms, more specific terms, and more general terms. If "jewelry box" isn't listed in the *Index*, try box or container. Or if the hinge is what makes it unique, look up hinge. Since most patents have more than one classification, chances are, the invention being researched will too. Look up all relevant aspects of the invention. Even if only part of it is covered in another patent, that part will not be new and therefore not patentable.

There are three versions of the *Index*, the printed version, available from the GPO, and two electronic versions. The CASSIS ASSIST DVD-ROM (found at PTDLs and many GPO Depositories) has a plain text version, and the USPTO Web site includes a (slightly hard to find) hypertext version which links to the Web version of the *Manual of Classification* (see Step Two). In all versions, there are two columns of numbers next to the terms indexed. The first number is the class and the second is the subclass. Some terms have only a class, others will have both numbers, and some will also have a plus (+). Write down all of what is there. Figure 2 shows a typical section from the Web version.[7]

Step Two: Manual of Classification

The *Manual of Classification* can be summed up as an outline of all patentable subject matter. The *Index* is the tool that points to the most likely spot in that outline. The first number from the *Index* is the class. Classes are the "chapters" in the manual. They are arranged in numerical order, with the single class for plant patents first, and the design

FIGURE 2. Sample Listing from the *Index*

```
Werner Complex Formation to Recover ........  585 / 850
            Hydrocarbon ..........................  585 / 850
Wet Suit ....................................    2 / 2.14+
Wetting (See Moistening)
            Compositions .........................  516 / 198+
            Textile spinning with ................   57 / 295
Whale Bone ..................................    2 / 255+
Whale Oil ...................................  554 / 1+
Wharves .....................................  405 / 284+
Wheel (See  Pulley; Wheeled)
     Aircraft
            Landing gear .........................  244 / 103 R+
            Paddle wheel propelled ...............  244 / 70
            Paddle wheel sustained ...............  244 / 19+
            Plane & paddle wheel sustained .....  244 / 9
                Heavier than air ...............  244 / 9
                Lighter than air ...............  244 / 27
            Prerotation ..........................  244 / 103 S
     Aligning tools ..........................   29 / 273
```

classes (which begin with "D" like the design patent numbers) after the utility classes.

The second number from the *Index* is the subclass. Subclasses may be in numerical order, but most classes have at least a few that are out of order. Figure 3 shows a sample from the Web version that illustrates this. Several subclasses will be in all capital letters. These are called main line subclasses. As mentioned previously, this is an outline of patentable subject matter. Subclasses indented under the main line subclass have one dot. Those indented under the one-dot subclass have two dots, and so on. Patents will only be classified in the most specific subclass that describes what is being patented. For those interested in this, more information about the classification system, and

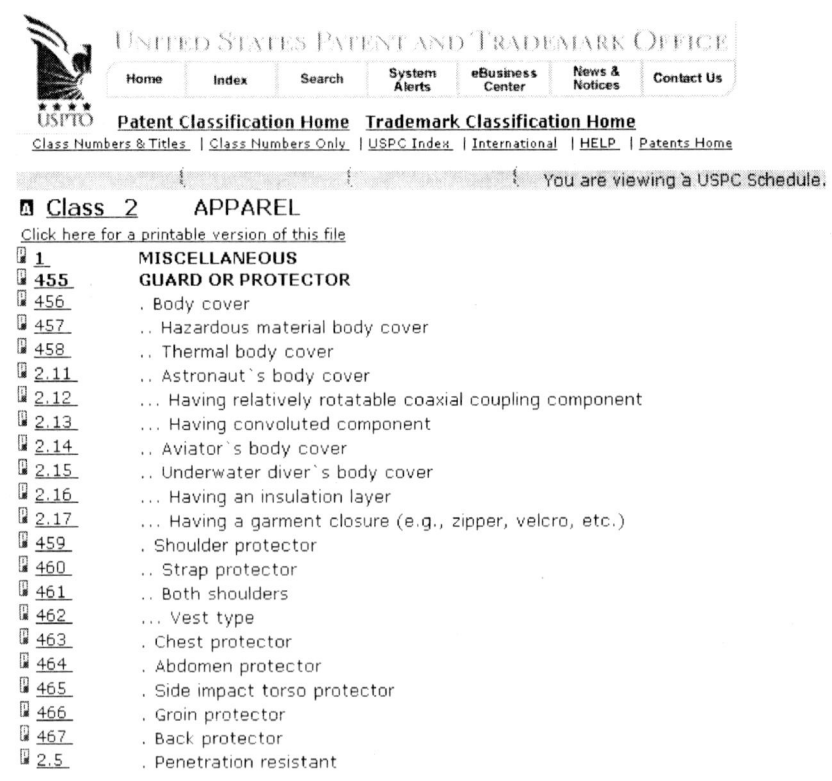

FIGURE 3. Sample from Online *Manual of Classification*

how classes and subclasses are selected for patents is available at the USPTO Web site.[8]

To find the subclass suggested, go to the first page of the class and scan the column until reaching the subclass. This will also give a general impression of what else is in the class. The best subclass may not be the one in the *Index*, so examine the surrounding subclasses or even the rest of the class to find the subclass(es) that best fit the invention.

If the *Index* listed a plus after the subclass, it is referring to a section of the outline–that subclass plus any indented under it. Examine them all (there may be two or dozens), and choose the most appropriate one for the invention. If no subclass was listed, there are possible subclasses throughout the class. Start at the beginning and browse the main line subclasses until an appropriate one is found, then browse the subclasses indented under it.

As with the *Index*, the *Manual* is available in both print and electronic formats. The print can be ordered through GPO. The electronic version has links to the next step, Definitions (click on the number), or to a search (click on the red P symbol).

Step Three: Classification Definitions

Most PTDLs have the definitions in microfiche. They are also available on the CASSIS disks and on the USPTO Web site. These are not necessarily dictionary definitions–many are self-referential. Their most useful features are the references to other subclasses, both within the same class and in other classes (see Figure 4).[9] Read the definition for the class as a whole (at the top or beginning), the subclass of interest, and any of its "parent" subclasses (under which it is indented). Is this an appropriate subclass for the invention being searched? What about the "See" references? Are any of them more suitable? Check the definitions for those classes/subclasses as well. It may not be fully clear that it is a good/bad classification until after some patents are examined.

TRANSLATING CLASSIFICATION INTO PATENTS

Many patrons mistakenly believe that the *Manual* is a list of patents. Actually each subclass has several patents in it, some have a hundred or more. When a subclass gets too full, the Classification Branch of the USPTO will conduct a reclassification project and create new sub-

FIGURE 4. Classification Definition

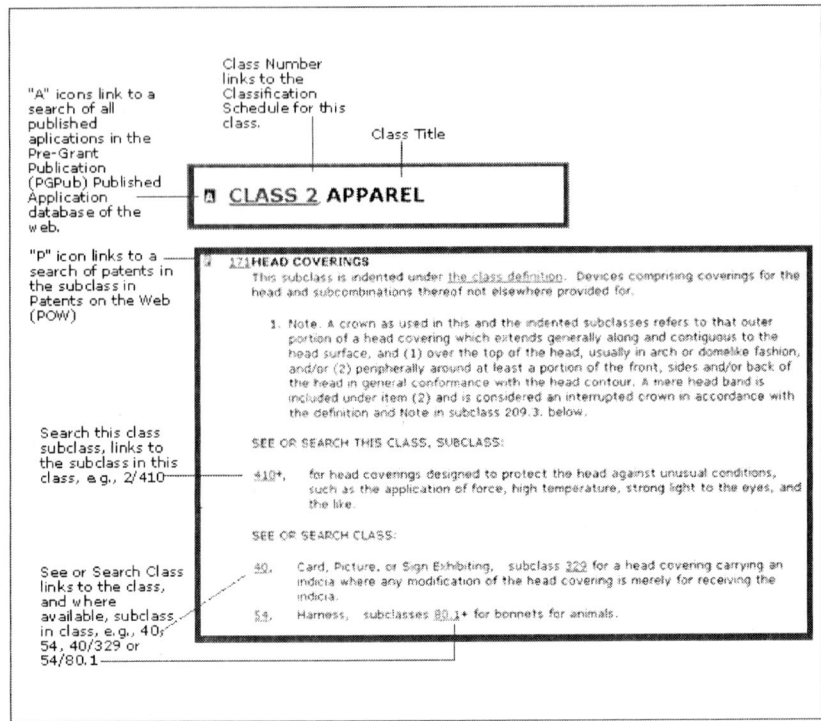

classes or even new classes. This is how those out of numerical order groups appear. Even the older patents will be reclassified (i.e., moved to the new classifications).

This is the part of the process where a computer must be used. Some libraries will offer to print the list of patent numbers (usually from CASSIS) when the patron determines the class/subclass.

Many U.S. patent databases (CASSIS and the USPTO's Web database among them) have a field for the current (i.e., updated) classification. This means that a search for a class/subclass will retrieve any patents ever granted by the USPTO that, if granted today, would be in that class/subclass. One main advantage of this in the USPTO's Issued Patents database (as well as CASSIS and the LEXIS/NEXIS Academic database) is that patents back to the beginning of the USPTO (1790) will be searched. A text search is limited by the text records in the database, usually only 1971 or 1976 forward.

Steps Four and Five: Databases

Any database (or combination of databases) that contains a complete collection of U.S. patents may be used. With one exception, this usually means first browsing the patent titles and abstracts in a text database (CASSIS Patents Bib or Lexis/Nexis Academic Law/Patents or any of the patent abstract databases), and then retrieving a list of patents in the subclass from a "numerical" database. Many beginning searchers find it useful to start with the text databases, even though the search is not complete, because the titles of the patents retrieved help validate the choice of subclass.

The numerical type (e.g., CASSIS Class or Lexis/Nexis Academic Law/Patents/Class) has two functions: search by class/subclass number to find all the patents in it, or search by patent number to find the current classification (all class(es)/subclass(es)).

In text databases, truncation is possible (to retrieve an entire class), as is limited subclass range searching (most ranges are literal so any of those out-of-order subclasses are not included). Classifications may be searched one at a time or "ORed" together. When the classification is uncertain, it is better to search one class/subclass at a time. This way, the patents can be reviewed for relevance, and an inappropriate classification discarded. In the numerical databases each class/subclass has its own record–a list of all the patents it contains.

Many searchers find it useful to go back and forth between the two types. For example, a searcher using CASSIS might browse titles/abstracts using the BIB disk until she is certain that the class/subclass is on target. Then she could switch to the Class disk to retrieve (print) the list of all patents classified in that class/subclass. And finally switch back to the BIB disk to select patents of further interest (patented since 1969).

The major exception to this dual search is the patent database at the USPTO Web site. A search by current classification will retrieve both full text files back to 1976 and earlier records that have no text, just a link to the page image file. The drawback with this database is in the display of results: the display (including the ability to print) is limited to fifty records at a time and, other than by laborious cut-and-paste, cannot be downloaded.

The point of the search is to throw a net wide enough to catch anything that could prevent a patent on the invention (or part of it). By examining any patent that could read on an invention (the prior art), inventors are prepared to explain why or how their invention differs or can design around any conflicts. Reviewing the prior art patents also helps inventors become familiar with "patentese," the special combination of legalese and scientific language that patents are written in. They are then better prepared to write the first draft of the application.

Step Six: Official Gazette

The patent section of *The Official Gazette of the United States Patent and Trademark Office* (known as the *OG*) is the weekly summary of patents granted. The publication date is identical to the patent issue date, and the patents are listed in numerical order. The *OG* began publication in 1871. For a short period of time it was published both in print and on CD-ROM. The print version was discontinued in September 2002.

Once a list of patent numbers has been generated, use the *OG* to view a brief description and a drawing for each patent number. If more detail is needed, move on to the next step. For searches carried out on databases that contain images, this intermediate step is not necessary. It is really for the convenience of searchers using collections of patent microfilm. Using the *OG* weeds out many patents, consequently fewer reels of microfilm will need to be retrieved, loaded, and viewed.

Step Seven: Patent Documents

Once patents of interest have been identified, the complete documents should be examined, including any drawings. Recent patents are only available in electronic form, but there are still several choices. The USPTO Web databases have TIFF[10] page images (and a link to a free viewer[11]). PTDLs and many GPO depositories receive the patents on a DVD-ROM file called USAPat. And there are fee-based full text patent databases that include the drawings.

For those who prefer microfilm, PTDLs are required to have at least a 20-year back file, and many have a complete collection on both microfilm and DVD. Some even have part of the collection in print! (Usually when a PTDL de-accessions bound patents, they find a home at another PTDL.)

The New Next Step

Now that the *OG* (Step 6) is not a part of searches done on the USPTO Web site, there is a new step that can be taken after patents have been examined–that of searching applications which have not yet granted as patents. The same Web page that provides access to the full text patent database also has a link to the pending applications database. As part of the harmonization of U.S. patent law with the rest of the world, U.S. patent applications are now (usually) published after 18 months. The database contains applications published since March 15, 2001.

Why should an inventor bother with this step? These applications are also considered prior art. If someone else has already applied for a patent on a similar invention, some or all of the searcher's invention may no longer be patentable. For the same reason, it is also a good idea to search other resources for anything that resembles the invention: technical journals, trade journals, catalogs such as Sweet's or the Thomas Register, trade organizations' Web sites, anywhere that would feature innovations.

WHAT IF NOTHING MATCHES?

The next question most inventors will ask is "What do I do if I cannot find anything like my invention?" If nothing even close is found, the classification is probably wrong. There are three basic alternatives.

1. Start over. Go back to the *Index, Manual,* and/or Definitions looking for other class/subclass numbers to try. Look up new terms in the *Index*. Look at the rest of the class (rather than just the subclasses that were searched). This is where the references in the definitions are helpful.
2. Try searching for words or phrases in the title, abstract or full text. When a related patent is found, record the classifications and look them up in the *Manual*. Go back to the database for a classification search for any that are relevant.
3. For those with more money than time: Hire a professional. Go to a patent agent or attorney (NOT an invention submission company). They must all pass a test (known colloquially as the patent bar) and remain in good standing with the USPTO. PDTLs have a published list, which is also on the USPTO Web site. Or hire a professional patent searcher. Many of these firms have been operating near the USPTO for generations, and are frequently hired by the law firms for searching.

Another option is to go to a PTDL. Although librarians at PTDLs cannot determine the classification for patrons, they have contact information for the Classification Branch of the USPTO that can suggest some appropriate classes and subclasses.

A NOTE ON CLASSIFICATION vs. KEYWORD SEARCHING

There is currently a debate as to the necessity of using the classification system, even to perform a search for prior art. Part of this debate is the same as the debate of controlled vocabulary vs. full text searching. The controlled vocabulary is just represented numerically with the U.S. Patent Classification System. An additional difficulty is that the text of many patents is written specifically so that they will not turn up in a text search. This keeps the technology from being exploited by others even after the patent expires. And most importantly, text searching in databases containing United States patents (or their abstracts) will rarely produce results before 1971. (A notable exception is the various databases from Chemical Abstracts, but even these generally stop in the 1960s for electrical and mechanical patents or the 1950s for chemical patents.) Only the classification or patent number can retrieve patents prior to 1976 in the USPTO's database.

For non-U.S. patents, U.S. classification is generally not available (although a few PTDLs have access to a special examiners' database that includes some non-U.S. patents by U.S. classification). The International Patent Classification (IPC) can be used instead, although since it is less specific, it can also be less useful. A text search of the database, or a search for a known patent in the technology (by number, inventor, etc.), will reveal the IPC to use.

A classification search is the only way to retrieve all (U.S.) prior art that the USPTO has deemed similar. Keyword (full text) searching is an approach that can help jump start classification searches where the *Index/Manual* approach is unproductive, but unless the searcher is familiar with all of the language of patents in the technology, it cannot replace a classification search. Once a similar patent has been found by means of a text search, record the classifications listed on the front page. Look these up in the *Manual*, and choose any that seem appropriate for further searching. Be sure to also examine the subclasses near by, or even throughout the class.

CONCLUSION

All libraries with Internet access now have a United States patent collection, which should be used to our patrons' advantage. Although complex and sometimes arcane, patent searching can yield results that are extremely useful. Inventors can be encouraged to move forward with their invention or to change it. Past innovations can be reviewed, either for avenues not taken or for competitive intelligence. This article has been limited to patents granted by the USPTO, but the scope of a search can be extended to multiple countries in the Esp@ce database.

NOTES

1. The USPTO's CASSIS (Classification and Search Support Information System) is provided to all PTDLs. It consists of six basic DVD-ROMs containing various types of patent and trademark data, and a series of DVD-ROMs containing facsimiles of patent and trademark documents. All are available for purchase from the USPTO <http://www.uspto.gov/web/offices/ac/ido/oeip/catalog/products/cassis.htm> (22 January 2003).

2. Library of Congress, "U.S. Copyright Office" <http://www.loc.gov/copyright> (13 September 2002).

3. Rimmer, Brenda M. *International Guide to Official Industrial Property Publications*. London: British Library, Science Reference and Information Service, 1988 (and 1990 and 1992).

4. 35 USC 101.

5. 35 USC 102.

6. "In the patent context, 'read on' means to literally describe. A patent is infringed if the patent's claims read on (literally describe) all elements of the infringing device." Elias, Stephen and Lisa Goldoftas. *Patent, Copyright & Trademark*. Berkeley: Nolo, 1999 (3rd ed.), p. 284.

7. United States Patent and Trademark Office, "Index to USPC" <http://www.uspto.gov/go/classification/uspcindex/indextouspc.htm> (13 September 2002). Click on the letter of the alphabet for a specific term.

8. United States Patent and Trademark Office, "How to use the Patent Classification System" <http://www.uspto.gov/web/offices/pac/dapp/sir/co/ovrvw.htm>.

9. This figure is taken from the Patent Classification Help Page at the USPTO Web site <http://www.uspto.gov/web/patents/classification/help.htm> (8 April 2003).

10. TIFF is an electronic image format with very high resolution, similar to GIF or JPEG, but with more detail.

11. AlternaTIFF. Other compatible plugins are listed at http://209.119.26.39/patft/images.htm.

APPENDIX. Patent Resources from the United States Patent and Trademark Office

Many documents produced by the USPTO are depository items, available for selection at GPO Depositories. If your library is a Regional Depository, you already have all of these. If your library is a Selective Depository, you may have them already, or you can talk to your documents librarian about adding them. Most of them are also available online (via http://www.uspto.gov).

Patent Search Tools

Item #	SuDoc Stem	Title	URL (or URL of Equivalent Information)
0262-A	C 21.9/2:	Attorneys and Agents Registered to Practice Before U.S. Patent Office (MF)	http://www.uspto.gov/web/offices/dcom/olia/oed/roster/index.html
0255-B	C 21.5/2:	Index of Patents (annual) (MF)	See Fulltext Databases
0255-A	C 21.5/2:	Index of Patents (annual) (P)	See Fulltext Databases
257	C 21.12/2:	Index to the U.S. Patent Classification (P)	http://www.uspto.gov/web/patents/classification/uspcindex/indextouspc.htm
0252-A	C 21.3/2:	List of Patent Classification Definitions (MF)	http://www.uspto.gov/web/patents/classification/selectnumwithtitle.htm
258	C 21.12:	Manual of Classification (P)	http://www.uspto.gov/web/patents/classification/selectnumwithtitle.htm
0260-B	C 21.5:	Official Gazette of United States Patent and Trademark Office: Patents (weekly) (MF)	See Notices or Fulltext Database
0260-A	C 21.5:	Official Gazette of United States Patent and Trademark Office: Patents (weekly) (P)	See Notices or Fulltext Database
0154-H-01	C 21.31/16:	Patents BIB and Patents SNAP (DVD)	See Fulltext Databases
0154-H-04	C 21.31/3:	Patents CLASS (DVD)	See Fulltext Databases
0252-A-01	C 21.3/3:	Reclassification Orders (series) (MF)	Not needed with online version
0260-E	C 21.31:	USAPat, Facsimile Images of U.S. Patents (DVD)	See Fulltext Databases

Trademark Search Tools

Item #	SuDoc Stem	Title	URL (or URL of Equivalent Information)
0256-D	C 21.5/3:	Index of Trademarks (annual) (MF)	See Fulltext Databases
0256-C	C 21.5/3:	Index of Trademarks (annual) (P)	See Fulltext Databases
0260-D	C 21.5/4:	Official Gazette of United States Patent and Trademark Office: Trademarks (weekly) (MF)	See Notices or Fulltext Database
0260-C	C 21.5/4:	Official Gazette of United States Patent and Trademark Office: Trademarks (weekly) (P)	See Notices or Fulltext Database
0154-B-11	C 21.31/9:	Trademarks ASSIGN (CD-ROM)	
0154-H-05	C 21.31/10:	Trademarks ASSIST (DVD)	
0154-H-03	C 21.31/14:	Trademarks REGISTERED and Trademarks PENDING (DVD-ROM) (aka Trademarks EIB)	See Fulltext Databases
0260-E-01	C 21.31/11:	USAMark: Facsimile Images of Registered United States Trademarks (CD-ROM)	See Fulltext Databases

Materials Relating to Patent Searching

Item #	SuDoc Stem	Title	URL (or URL of Equivalent Information)
260	C 21.5/5:	Consolidated Listing of Official Gazette Notices Re Patent and Trademark Office Practices and Procedures, Patent Notices (annual) (P)	http://www.uspto.gov/web/offices/com/sol/og/index.html
0256-A-02	C 21.26/2:	General Information Concerning Patents (P)	http://www.uspto.gov/web/offices/pac/doc/general/index.html
0256-A-01	C 21.26:	General Information Concerning Trademarks (P)	http://www.uspto.gov/web/offices/tac/doc/basic/
259	C 21.15:	Manual of Patent Examining Procedures (P)	http://www.uspto.gov/web/offices/pac/mpep/index.html
0154-H-02	C 21.31/13:	Patents ASSIGN and Trademarks ASSIGN (DVD)	
0154-H-06	C 21.31/5:	Patents ASSIST (DVD-ROM)	
		USPTO (and PCT) forms	http://www.uspto.gov/web/forms/index.html
		USPTO fees	http://www.uspto.gov/web/offices/ac/qs/ope/fees.htm

37

APPENDIX (continued)

Other USPTO Documents

Item #	SuDoc Stem	Title	URL (or URL of Equivalent Information)
0260-E-02	C 21.31/12:	Cassis Currents Newsletter (EL)	
0188-A-10	C 51.16:	Catalog of Government Patents (P)	
0254-B	C 21.28:	Directories (MF)	
0260-E-02	C 21.32:	Electronic Products (misc.) (E)	
254	C 21.2:	General Publications	
0254-A	C 21.14/2:	Handbooks, Manuals, Guides	
0254-B-02	C 21.28/2:	Information Directory (annual) (MF)	http://www.uspto.gov/main/contacts.htm
		Patent and Trademark Depository Library Program	http://www.uspto.gov/web/offices/ac/ido/ptdl/index.html
260	C 21.5/4-2:	Patent and Trademark Office Notices (annual) (P)	http://www.uspto.gov/web/offices/com/sol/og/index.html
260	C 21.5/4 A:	Patent and Trademark Office Notices (weekly) (P)	http://www.uspto.gov/web/offices/com/sol/og/index.html
251	C 21.1/2:	Patent and Trademark Office Review (annual) (MF)	
261	C 21.7:	Patent Laws (P)	http://www.uspto.gov/web/offices/pac/mpep/mpep.htm
0254-B-01	C 21.30:	Products and Services Catalog (annual) (P)	http://www.uspto.gov/web/offices/ac/ido/oeip/catalog/index.html
0260-E-02	C 21.33:	PTO Pulse (weekly w/monthly cumm.) (EL)	
262	C 21.14:	Regulations, Rules, Instructions (P)	

38

Finding Grandpa's Patent: Using Patent Information for Historical or Genealogical Research

Jan Comfort

SUMMARY. Hardly a week goes by without a phone call, an e-mail message, or a visit from a patron who is looking for a patent issued to a family member. Often, the patron does not have complete information, and sometimes has nothing more than a name and a notion that Grandpa once invented something. This article describes some strategies that can be used by Librarians to assist patrons in historical patent research. It, additionally, includes a list of sources and highlights special materials that are available at Patent and Trademark Depository Libraries. *[Article copies available for a fee from The Haworth Document Delivery Service: 1-800-HAWORTH. E-mail address: <docdelivery@haworthpress.com> Website: <http://www.HaworthPress.com> © 2001 by The Haworth Press, Inc. All rights reserved.]*

KEYWORDS. Inventors–historic, inventions–historic, historic United States patents, patent research, genealogical research

---

Jan Comfort is Government Documents Reference Librarian, R. M. Cooper Library, Clemson University, Box 343001, Clemson, SC 29634-3001 (E-mail: comforj@clemson.edu).

[Haworth co-indexing entry note]: "Finding Grandpa's Patent: Using Patent Information for Historical or Genealogical Research." Comfort, Jan. Co-published simultaneously in *Science & Technology Libraries* (The Haworth Information Press, an imprint of The Haworth Press, Inc.) Vol. 22, No. 1/2, 2001, pp. 39-56; and: *Patent and Trademark Information: Uses and Perspectives* (ed: Virginia Baldwin) The Haworth Information Press, an imprint of The Haworth Press, Inc., 2001, pp. 39-56. Single or multiple copies of this article are available for a fee from The Haworth Document Delivery Service [1-800-HAWORTH, 9:00 a.m. - 5:00 p.m. (EST). E-mail address: docdelivery@haworthpress.com].

http://www.haworthpress.com/store/product.asp?sku=J122
© 2001 by The Haworth Press, Inc. All rights reserved.

INTRODUCTION:
BUT IT'S ALL ON THE INTERNET, RIGHT?

One of the most important things to remember when beginning a search for historical patents is that not everything is available in electronic format. The United States Patent and Trademark Office (USPTO) maintains a database of every patent issued, from 1790 to the present. It is available at http://www.uspto.gov/patft/index.html. There are some limitations to keep in mind, however. Patents from 1790 to 1975 are only searchable by patent number and current U.S. Classification. For these patents the typical keyword or inventor's name search is not possible. However, the searcher with only Grandpa's name generally does not know the patent number or the current U.S. Classification. Most often all they know is the name of the inventor, where they lived, and a range of years that they might have been inventing. Finding the patent number with just this information can be challenging.

Determining the current classification of a historic patent can be equally as challenging. All patents have been assigned to a class (or an even more specific class and subclass) based on the use or function of the invention. A classic example is that cement mixers and washing machines could both be assigned to class 209 (Classifying, Separating, and Assorting Solids) because they both agitate. Determining the class and subclass is not always a simple thing. Very few sources of information about older inventions list classification information. Those that do will list the class and subclass to which the invention was assigned at the time of issue. If the patent in question is more than a few years old, it is conceivable that the class and subclass numbers have changed. They may even have changed more than once. Every time such changes are made, the information is recorded in the patent database, and only the current classification information is retained. Clearly it is going to take some investigating to uncover Grandpa's patent.

THE GOAL OF EVERY PATENT SEARCH

Regardless of what information is provided, the goal of every patent search is the same: to determine the patent number. When the patent number is found, so is the patent. But how does one find the patent number? Like most library research projects, the simplest way is to consult an index or a published list. The bibliography for this article lists many such indexes. It is arranged by access point, so with only the name of the inventor, or the subject matter of the invention, the searcher can check the sources that are listed in the appropriate section. Those two sections are arranged with the most useful sources listed first. Every index will lead to the patent number, along with other useful information.

Working with patent numbers can be challenging as well. The first United States patent was issued on July 31, 1790. However, the early Patent Board did not think to assign numbers to the patents. It was not until July 13, 1836 that patents began to be numbered. Patents issued before this date are often referred to as Name and Date patents. As a further complication, on December 15, 1836 there was a disastrous fire at the Patent Office, and all of the patents–nearly 10,000–were destroyed. Luckily, a list of issued patents survived, and at this point the Name and Date patents were numbered. To avoid confusion, the numbers were preceded with an X (i.e., the first patent ever issued is numbered X1). Many efforts were made to recover the lost patents, and as copies were received, they were recorded, so there is information about many of these patents in the patent searching database at the USPTO home page.

WHAT DO I DO IF I DON'T FIND ANYTHING?

But what if you don't find anything? Again, some traditional library research methods might prove to be helpful.

- *Try alternate spellings.* In the 18th and 19th centuries, people were much less concerned about proper or standardized spelling, even of surnames.
- *Expand the dates you are searching.* Life spans were shorter, and people were more likely to work up until the time of their death. If you know that a person lived from 1820-1865, search indexes from 1835-1870. (Occasionally patent rights were assigned to surviving family members.)
- *Try to determine if grandpa worked with another inventor, or with a company.* Sometimes a patent is assigned to a company. Perhaps the inventor worked for this company and his work on the invention was part of his job, so his name may or may not be listed as inventor. Or perhaps the inventor sold (assigned) his patent rights to a company or another individual. If Grandpa assigned these rights at the time that the patent issued, Grandpa's name should be listed in the Index of Patents, along with a cross-reference to the name of

the company or individual. Because of the way the indexes are formatted, sometimes the assignee name is easier to spot.
- *Could it have been a patent issued by a state or a colony?* Before the United States Patent system was formally organized, many of the colonies had their own systems in place. Limited information exists about the patents issued to colonies. However, many states continue to issue patents–with rights that extend only within the borders of their state. There is no central database of these patents, and each state handles things in a different way. Often the Department of State issues these patents. Contact the Department of State of the state in question for information.
- *Could it have been a patent issued by another country?* The United States of America is a relatively new country, made up of people who came from many places. Perhaps Grandpa patented his invention in the old country. Many different countries issue patents, and the amount of available information on older patents varies greatly. See the bibliography for several sources of information.
- *Was the invention ever patented?* There are many reasons why an invention may never have been patented. Even in the early days, getting a patent was expensive and time-consuming, and often involved trips to lawyers or to Washington, D.C. This may have put it out of inventor Grandpa's reach. It is also entirely possible that Grandpa did invent something, but he just used it himself, or made several to share or sell on a limited basis. Unfortunately, this would probably be impossible to verify without further information.
- *But I have an object with a patent number (or "Patent Pending") stamped on it!* Sometimes what looks like a patent number is not a patent number. It could be any number of things–a manufacturer's model number, for example. A good way to verify the patent number is to consult the list of patent numbers by year from the USPTO Web page: http://www.uspto.gov/web/offices/ac/ido/oeip/taf/issuyear.htm. Keep in mind that decorative objects could have been protected by design patents, so pay close attention to those patent numbers that start with a "D." If an object is stamped with "Patent Pending," there are several reasons why a patent may never have been granted. It is possible that it was applied for but rejected for some reason. Maybe the company merged with another company, or shut down altogether. Or perhaps the individual or company never got around to applying for the patent. There were fewer laws about these kinds of things in the old days, and it was apparently fairly common to use "Patent Pending" or even bogus patent numbers.

OTHER SOURCES OF INFORMATION

If you have gone through all of the previous steps and have come up empty, there are still a few things that you (or your patron) can do. Consult local or county histories. If an inventor or invention played a prominent role, it is possible that either would have been included in a published history of the area. It might be a little difficult to uncover these sources, especially if the inventor is from a different state. At this point the search might be extended to OCLC (Online Computer Library Center) or another global catalog of library holdings, to an Internet search engine or database such as *Amazon.com* or *Google*, and/or to on-line catalogs for large libraries in the area. Keep in mind that many older or specialized sources, especially government documents, might not be listed in a library's online catalog. If a patron can identify a particular book, Interlibrary Loan could be an option. Local historical societies may also be helpful. Many of them have book collections which may or may not be publicly accessible.

Consult a book about the specific area of technology. There are so many collectors out there these days, and for every type of collection there are books. Consult a book on the history of eggbeaters, and you may just find the information that you need.

CONCLUSION

There are as many strategies for researching historical patents as there are historical patents. The Bibliography that follows lists a few of the most important resources that can be the basis of a successful search. But perhaps the best strategy of all is to consult your local Patent and Trademark Depository Library (PTDL) Librarian. Not only do these librarians have access to a wealth of printed and electronic tools; they are experts in patent searching. Whether using old printed indexes that are crumbling into dust, or the latest update on the USPTO Web page, PTDL Librarians are committed to excellent service, and most also have a deep interest in history and the persistence to work on a search until something is found. There is at least one PTDL in every state. (See Appendix or http://www.uspto.gov/go/ptdl/ptdlib_1.html for a complete list.) Call ahead before referring a patron, as collections vary from library to library, and each has its own policies. A phone call can verify library holdings, special locations or limited hours, and the level of assistance available. Sometimes the PTDL Librarian can even locate the patent and either send it or explain how it can be obtained. With a little bit of luck and a little persistence, you can preserve history and make your patron's day.

BIBLIOGRAPHY

HISTORICAL INVENTORS: A SELECTED BIBLIOGRAPHY OF SOURCES

1. Inventors Listed by Subject Matter of Invention

Books and Articles

Subject Matter Index of Patents for Inventions Issued by the United States Patent Office from 1790 to 1873, Inclusive. [Reprint.] Series Title: America in Two Centuries, An Inventory. North Stratford, NH: Ayer Company Publishers, Incorporated, 2000.

This is one of several reprints of the 1874 ed. published by the Government Printing Office, Washington D.C. (SuDoc no. I 23.7: 874/1-3. See Alphabetic List, Below). It is a three-volume set, arranged alphabetically by subject of invention. Each entry lists inventor's name, inventor's city and state of residence, date of patent grant, and patent number. Unfortunately, there is no index by inventor's last name. However, this is still the very best source for identifying historic inventors and inventions, and it is available at nearly every PTDL, as well as many large public, academic, or special libraries. It is also included as a part of the 19th Century Masterfile Database, see entry under databases, below.

United States Patent Office. *Annual Report of the Commissioner of Patents.* Washington, D.C.: Government Printing Office. 1843-1919.

Also known as the Patent Office Report, the arrangement of this report varies from year to year. Early reports were also included in the Congressional Serial Set. It usually consists of two parts: Agriculture and Mechanics. From 1853-1868, the Mechanics part was published in two volumes. The first volume contains a list of patents by subject and patentee, and the second volume contains very brief abstracts of the patents: about 5-10 abstracts per page. This volume is arranged by subject of invention from 1854-1858, and numerically for the remaining years. The Annual Report was published until 1965, but in 1920 the index portion began being published under the title of Index of Patents Issued from the United States Patent and Trademark Office (see below). In addition to the indexes, the front matter of this report contains a wealth of interesting historical information, such as number of patent applications filed and patents granted, and number of patents issued in foreign countries.

United States Patent Office. *Index of Patents Issued from the United States Patent and Office.* Part II: Index to Subject of Inventions. Washington, D.C.: Government Printing Office.

The Index of Patents is a cumulative index that lists patents granted in a given year. This section is arranged alphabetically by class and subclass. Entries also list the inventor's name and any assignment information that was known at the time of issue, patent title, patent number, and date of issue. From 1843-1919 the index is actually part of the Annual Report of the Commissioner of Patents (also known as the Patent Office Report). Often it is bound separately–particularly from 1872, when it began listing references to volume and page numbers of the Official Gazette. (See next entry.) Beginning in 1966, the two sections of this publication are bound in two separate volumes–Part I (Index to Patentees), and Part II (Index to Subject of Inventions).

United States Patent and Office. *Official Gazette of the United States Patent Office.* Washington, D.C.: Government Printing Office. 1872-

The Official Gazette has been printed every Tuesday since 1872. Numerical entries are organized by class and subclass before they are numbered. Therefore this source has a general subject arrangement. Each issue has two indexes–one an alphabetical list of inventions, and the other an alphabetic list of inventors. Index entries also contain the patent number, date of issue, inventor's city and state of residence, and any assignment information that was known. The major drawback of this publication is that unless you have an exact date to go by, it involves a lot of searching, since it is issued weekly. There is some monthly and quarterly indexing, but not consistently. Beginning with May 18, 1965 there is also an index listing patent numbers by state. Unfortunately, this geographic index is not picked up in the annual Index of Patents. The

Official Gazette contains bibliographic information, a drawing (if one was submitted), and at least one claim reprinted from the patent document.

2. Alphabetic Lists of Inventors

Books and Articles

United States Patent and Trademark Office. *Index of Patents Issued from the United States Patent and Trademark Office.* Part I: List of Patentees. Washington, D.C.: Government Printing Office.

This section is arranged alphabetically by name of inventor (and assignee, if that was known at the time of issue). Entries also list the patent title, patent number, date of issue, and class/subclass. Beginning in 1872, there are also references to the volume and page number in the Official Gazette, where drawings and specifications can be found.

United States Patent Office. *List of Patents for Inventions and Designs, Issued by the United States from 1790-1847, with the Patent Laws and Notes of Decisions of the Courts of the United States for the Same Period: Compiled Under the Direction of Edmund Burke, Commissioner of Patents.* Washington, D.C.: J & S Gideon, 1847.

United States Patent Office. *A Digest of Patents Issued by the United States, from 1790-January 1, 1839, Published by Act of Congress Under the Superintendence of the Commissioner of Patents, Henry L. Ellsworth, To Which is Added the Present Law Relating to Patents.* Washington, D.C.: Peter Force, 1840.

United States Congress. *Letter from the Secretary of State Transmitting a List of All Patents Granted by the United States, The Acts of Congress Relating Thereto, and the Decisions of Courts of the United States Under the Same.* Document No. 50, 21st Congress, 2d Session, January 13, 1831. Serial Set No. 207. Washington, D.C.: Duff Green, 1831.

Each of these three reports arranges patents alphabetically by "class," although the classes are not numbered as is customary, so it is actually a subject arrangement. At the end of each of the "class" sections there is an index by inventor's name. Because of this, these original reports can be more useful than the reprints.

The New American State Papers 1789-1860: Science and Technology. Prof. Thomas C. Cochran, ed. Wilmington, DE: Scholarly Resources, Inc., 1973.

This is a fourteen volume series drawing together materials from the original American State Papers (published between 1832 and 1861); U.S. Congressional Serial Set volumes (published continuously since 1817); and documents from the Legislative Section of the National Archives. Volumes 4 and 5 relate to patents, and reprint many interesting reports. In particular, volume 4 has a reprint of Document No. 50 (see above).

3. Inventors Listed by Geographic Region

Books and Articles

Calkin, Homer L. and Corrine Calkin. "Iowa Inventors and Inventions Part One." *The Palimpsest* L (1969): 369-432.

Calkin, Homer L. and Corrine Calkin. "Iowa Inventors and Inventions 1843-1873 Part Two." *The Palimpsest* L (1969): 433-481.

Comfort, Jan. "South Carolina Inventors and Inventions 1790-1873." *The South Carolina Magazine of Ancestral Research* XXV (1997): 123-136.

Fulghum, R. Neil. *North Carolina Patents, 1790-1873.* Raleigh, North Carolina: North Carolina Museum of History, 1979.

Holsclaw, Birdie. "Early Colorado Inventors: Colorado Patent Models in the Cliff Peterson Collection 1852-1890." *The Colorado Genealogist* 49 (1988): 34-36.

Marhenke, Chris. "Florida Inventors." [Electronic Resource]. Not yet available to the public. (Printouts distributed to Florida PTDLs.) 2002.

Sharrer, Terry G. "Patents by Marylanders, 1790-1830." *Maryland Historical Magazine* 71(1976): 50-59.

Walston, Mark. "Maryland Inventors and Inventions 1830-1860." *Maryland Historical Magazine* 80 (1985): 66-93.

"Wisconsin Inventors." *Badger History (Special Issue).* 23 (1970): 2-64, plus Teacher's Supplement.

Woodcroft, Bennet. *Alphabetical Index of Patentees of Inventions: from March 2, 1617 (14 James I) to October 1, 1852 (16 Victoriae).* New York: Augustus M. Kelley, 1969.

Reprint of 1854 edition covering British Patents

4. Electronic Databases

(Can be searched by inventor's name, geographic region, or subject matter of invention)

Carnegie Library of Pittsburgh. *From Air Brakes to Zinc Furnaces: Pittsburgh and Allegheny City Patentees* 1790-1879. [Electronic Resource]. Retrieved June 28, 2002, from http://www.clpgh.org/clp/Scitech/invent/.

The European Patent Office. *Espacenet.* [Electronic Resource]. Retrieved February 20, 3003, from http://ep.espacenet.com/.

Search by keyword from the opening screen. Or click on the link "Worldwide–30 million documents" to reach the search screen that permits searching by inventor, date, or classification. This database does not have complete coverage of older patents. A search of the dates of coverage shows that the earliest U.S. patent included is number 23,045 dated February 22, 1859.

Hocker, Susan E. Miami University Library. *Index to Early Louisiana Patents, 1810-1890.* [Electronic Resource]. Retrieved February 20, 2003, from http://adler.lib.muohio.edu/~shocker/LAPAT/index.html.

Great Lakes Patent and Trademark Center. Detroit Public Library. *African-American Inventors Database.* [Electronic Resource]. Retrieved June 28, 2002 from http://www.detroit.lib.mi.us/glptc/aaid/index.asp.

State Library of Iowa. *Iowa Inventions Database.* [Electronic Resource]. Retrieved February 20, 2003, from http://www.silo.lib.ia.us/app/cgi-bin/patents/.

Subject Matter Index of Patents Issued by the U.S. (1790-1873). [Electronic Resource]. Patent and Government Document Indices. 19th Century Masterfile. Reston, VA: Paratext, Inc., 1999-

Wyoming State Library. *Wyoming Inventors Database.* [Electronic Resource]. Retrieved June 28, 2002, from http://cowgirl.state.wy.us/inventors/.

5. Special Materials Available at Patent and Trademark Depository Libraries

Books and Articles

Dobyns, Kenneth W. *The Patent Office Pony: A History of the Early Patent Office.* Fredericksburg, VA: Sergeant Kirkland's Museum and Historical Society, 1997.

This is a terrific book about the history of the Patent Office. In this context, of particular note is a list of all known patents issued to the Confederate Patent Office on p. 206-216.

Randall, Merle, and Evelyn Boelter Watson. *Finding List for United States Patent, Design, Trademark, Reissue, Label, Print, and Plant Patent Numbers.* Berkeley, CA: University of California Press, 1938.

United States Department of Commerce. Patent and Trademark Office. *Name and Date Patents July 31, 1790-July 2, 1836.* Washington, D.C.: Patent and Trademark Depository Library Program, August 19, 1999.

This is actually a photocopy of a typed (and in some cases handwritten) card file used in the Public Search Room at the United States Patent and Trademark Office. Before the publication of the Subject Matter Index, it was the only access to the early unnumbered patents.

United States Department of Commerce. Patent and Trademark Office. *Patent and Trademark Office Collection of Historical and Interesting U.S. Patents in Celebration of Our Nation's Bicentennial. Microfilm Index.* Indexed by Bruce B. Cox and Amy K. England. Washington, D.C.: Patent and Trademark Depository Library Program, April 1987.

United States. National Archives and Records Service. *Additional Improvement Patents, 1837-1861*. Special List no. 39. Compiled by James A. Paulauskas. Washington, D.C.: General Services Administration, 1977.

United States. National Archives and Records Service. *Preliminary Inventory of the Records of the Patent Office (Record Group 41)*. Compiled by Forrest R. Holdcamper. Washington, D.C.: National Archives Editorial Division, November 1966.

United States Patent Office. "Historical Notices of Inventions from Archives of the States." *Report of the Commissioner of Patents for the Year 1849, Part 1: Arts and Manufactures*. Washington, D.C.: Office of Printers to the Senate, 1850.

United States Patent Office. "Papers and Abstracts Relating to Early American Inventors from the Archives of States." *Report of the Commissioner of Patents for the Year 1850, Part 1: Arts and Manufactures*. Washington, D.C.: Office of Printers to House of Reps., 1851.

Women Inventors

United States Department of Labor. Women's Bureau. *Women's Contributions in the Field of Invention: A Study of the Records of the United States Patent Office*. Bulletin of the Women's Bureau, No. 28. Washington, D.C.: Government Printing Office, 1923.

United States Patent and Trademark Office. Technology Assessment and Forecast Program. *Buttons to Biotech 1996 Update Report: U.S. Patenting by Women, 1977 to 1996*. Washington, D.C.: United States Patent and Trademark Office, 1998.

United States Patent Office. *Women Inventors to whom Patents have been Granted by the United States Government 1790-July 1, 1888*. Compiled under the Direction of the Commissioner of Patents. Washington, D.C.: Government Printing Office, 1888.

United States Patent Office. *Women Inventors to whom Patents have been Granted by the United States Government: July 1, 1888 to October 1, 1892: Appendix No. 1*. Compiled under the Direction of the Commissioner of Patents. Washington, D.C.: Government Printing Office, 1892.

United States Patent Office. *Women Inventors to whom Patents have been Granted by the United States Government: October 1, 1892 to March 1, 1895: Arranged Chronologically and by Classes. Appendix No. 2*. Compiled under the Direction of the Commissioner of Patents. Washington: Government Printing Office, 1895.

Electronic Resources

USAPat: Facsimile Images of United States Patents Issued 1790 to Present. [Electronic Resource]. Washington, D.C.: United States Patent and Trademark Office. Office for Patent and Trademark Information. 2000-

The same patents that are published on the USPTO Web site are also provided to PTDLs on DVD-ROM. They can be printed much more quickly than the patents on the Internet, especially using the heavy-duty printer provided by the Patent Office. They are also available as a selection for Federal Depository Libraries.

CASSIS: Classification Information Files. [Electronic Resource]. Washington, D.C.: United States Patent and Trademark Office. Office of Electronic Information Products and Services.

This file contains class and subclass information on every United States Patent issued since 1790. It is only accessible by patent number. This is just one of a series of optical disk products called CASSIS, which are provided to all Patent and Trademark Depository Libraries.

APPENDIX. List of Patent and Trademark Depository Libraries, Alphabetical by State

City and State	Library Name	Phone Number
Alabama		
Auburn	Ralph Brown Draghon Library, Auburn University	(334) 844-1737
Birmingham	Birmingham Public Library	(205) 226-3620
Alaska		
Anchorage	Z. J. Loussac Public Library, Anchorage Municipal Libraries	(907) 562-3620
Arizona		
Tempe	Noble Science and Engineering Library, Arizona State University	(480) 965-7010
Arkansas		
Little Rock	Arkansas State Library	(501) 682-2053
California		
Los Angeles	Los Angeles Public Library	(213) 228-7220
Sacramento	California State Library, Library–Courts Building	(916) 654-0069
San Diego	San Diego Public Library	(619) 236-5813
San Francisco	San Francisco Public Library	(415) 557-4500
Sunnyvale	Sunnyvale Center for Innovation, Invention, and Ideas (SCI^3)	(408) 730-7300
Colorado		
Denver	Denver Public Library	(720) 865-1711
Connecticut		
Hartford	Hartford Public Library	(860) 543-8628
New Haven	New Haven Free Public Library	(203) 946-7452
Delaware		
Newark	University of Delaware Library	(302) 831-2965
District of Columbia		
Washington	Founders Library, Howard University	(202) 806-7252

City and State	Library Name	Phone Number
Florida		
Ft. Lauderdale	Broward County Main Library	(954) 357-7444
Miami	Miami-Dade Public Library	(305) 375-2665
Orlando	University of Central Florida Libraries	(407) 823-2562
Tampa	University of South Florida, Tampa Campus	(813) 974-2726
Georgia		
Atlanta	Georgia Institute of Technology	(404) 894-4508
Hawaii		
Honolulu	Hawaii State Library	(808) 586-3477
Idaho		
Moscow	University of Idaho Library	(208) 885-6235
Illinois		
Chicago	Chicago Public Library	(312) 747-4450
Springfield	Illinois State Library	(217) 782-5659
Indiana		
Indianapolis	Indianapolis-Marion County Public Library	(317) 269-1741
West Lafayette	Siegesmund Engineering Library, Purdue University	(765) 494-2872
Iowa		
Des Moines	State Library of Iowa	(515) 242-6541
Kansas		
Wichita	Ablah Library, Wichita State University	(316) 978-3155
Kentucky		
Louisville	Louisville Free Public Library	(502) 574-1611
Louisiana		
Baton Rouge	Troy H. Middleton Library, Louisiana State University	(225) 388-8875
Maine		
Orono	Raymond H. Fogler Library, University of Maine	(207) 581-1678
Maryland		
College Park	Engineering and Physical Sciences Library, University of Maryland	(301) 405-9157

APPENDIX (continued)

City and State	Library Name	Phone Number
Massachusetts		
Amherst	Physical Sciences and Engineering Library	(413) 545-1370
Boston	Boston Public Library	(617) 536-5400, ext. 2265
Michigan		
Ann Arbor	Media Union Library, University of Michigan	(734) 647-5735
Big Rapids	Abigail S. Timme Library, Ferris State University	(231) 591-3500
Detroit	Great Lakes Patent and Trademark Center, Detroit Public Library	(313) 833-3379
Minnesota		
Minneapolis	Minneapolis Public Library	(612) 630-6120
Mississippi		
Jackson	Mississippi Library Commission	(601) 961-4111
Missouri		
Kansas City	Linda Hall Library	(816) 363-4600
St. Louis	St. Louis Public Library	(314) 241-2288, ext. 390
Montana		
Butte	Montana Tech Library, University of Montana	(406) 496-4281
Nebraska		
Lincoln	Engineering Library, University of Nebraska–Lincoln	(402) 472-3411
Nevada		
Las Vegas	Clark County Library, Las Vegas	(702) 733-1165
Reno	University of Nevada	(775) 784-6500, ext. 257
New Hampshire		
Concord	New Hampshire State Library	(603) 271-2239
New Jersey		
Newark	Newark Public Library	(973) 733-7779
Piscataway	Library of Science and Medicine, Rutgers University	(732) 445-2895

City and State	Library Name	Phone Number
New Mexico		
Albuquerque	Centennial Science and Engineering Library, University of New Mexico	(505) 277-4412
New York		
Albany	New York State Library	(518) 474-5355
Buffalo	Buffalo and Erie County Public Library	(716) 858-7101
Rochester	Central Library of Rochester and Monroe County	(716) 428-8110
Stony Brook	Melville Library, SUNY at Stony Brook	(631) 632-7148
North Carolina		
Raleigh	D.H. Hill Library, North Carolina State University	(919) 515-2935
North Dakota		
Grand Forks	Chester Fritz Library, University of North Dakota	(701) 777-4888
Ohio		
Akron	Akron-Summit County Public Library	(330) 643-9075
Cincinnati	The Public Library of Cincinnati	(513) 369-6971
Cleveland	Cleveland Public Library	(216) 623-2870
Columbus	Ohio State University	(614) 292-3022
Dayton	Wright State University	(937) 775-3521
Toledo	Toledo/Lucas County Public Library	(419) 259-5209
Oklahoma		
Stillwater	Oklahoma State University	(405) 744-7086
Oregon		
Portland	Lewis & Clark College	(503) 768-6786
Pennsylvania		
Philadelphia	The Free Library of Philadelphia	(215) 686-5331
Pittsburgh	The Carnegie Library of Pittsburgh	(412) 622-3138
University Park	Business Library, Paterno Library, The Pennsylvania State University	(814) 865-6369
Puerto Rico		
Bayamon	General Library, Bayamon Campus, University of Puerto Rico	(787) 786-5225
Mayaguez	General Library, Mayaguez Campus University of Puerto Rico	(787) 832-4040, ext. 2022

APPENDIX (continued)

City and State	Library Name	Phone Number
Rhode Island		
Providence	Providence Public Library	(401) 455-8027
South Carolina		
Clemson	R.M. Cooper Library, Clemson University	(864) 656-3024
South Dakota		
Rapid City	Devereaux Library, South Dakota School of Mines and Technology	(605) 394-1275
Tennessee		
Nashville	Stevenson Science and Engineering Library, Vanderbilt University	(615) 322-2717
Texas		
Austin	McKinney Engineering Library, The University of Texas at Austin	(512) 495-4500
College Station	Texas A&M University	(979) 845-5745
Houston	Fondren Library, Rice University	(713) 348-5483
Lubbock	Texas Tech University Library	(806) 742-2282
San Antonio	San Antonio Public Library	(210) 207-2500
Utah		
Salt Lake City	Marriott Library, University of Utah	(801) 581-8394
Vermont		
Burlington	Bailey/Howe Library, University of Vermont	(802) 656-2542
Virginia		
Richmond	James Branch Cabell Library, Virginia Commonwealth University	(804) 828-1104
Washington		
Seattle	Engineering Library, University of Washington	(206) 543-0740
West Virginia		
Morgantown	Evansdale Library, West Virginia University	(304) 293-4695, ext. 5113
Wisconsin		
Madison	Kurt F. Wendt Library, University of Wisconsin	(608) 262-6845
Milwaukee	Milwaukee Public Library	(414) 286-3051
Wyoming		
Cheyenne	Wyoming State Library	(307) 777-7281

esp@cenet®:
Europe's Network of Patent Databases

Gerry McKiernan

SUMMARY. *esp@cenet®* is a free Web-based patent search and document service provided by the European Patent Office (EPO). It offers desktop access to the bibliographic data and full text of patent applications and granted patents of the EPO, European national patent agencies, the World Intellectual Patent Organization, and many non-European nations, including the United States and Japan. As of early 2003, *esp@cenet®* provided access to more than 42.5 million patent documents for more than 70 countries. This article reviews the variety of basic and advanced interfaces and gateways within *esp@cenet®*, as well as its search, browse and retrieval options, and other components. For librarians, as well as non-specialists, *esp@cenet®* offers a central framework that provides comprehensive access to a wide range of significant national, regional, and international patent collections. *[Article copies available for a fee from The Haworth Document Delivery Service: 1-800-HAWORTH. E-mail address: <docdelivery@haworthpress.com> Website: <http:// www.HaworthPress.com> © 2001 by The Haworth Press, Inc. All rights reserved.]*

Gerry McKiernan, AB, MS, is Associate Professor, and Science and Technology Librarian and Bibliographer, Iowa State University, Ames, IA.

The author wishes to acknowledge the support and assistance provided by members of the European Patent Office in preparing this review and for permission to reproduce selected screen images from *esp@cenet®* and *epoline®*.

[Haworth co-indexing entry note]: "*esp@cenet®*: Europe's Network of Patent Databases." McKiernan, Gerry. Co-published simultaneously in *Science & Technology Libraries* (The Haworth Information Press, an imprint of The Haworth Press, Inc.) Vol. 22, No. 1/2, 2001, pp. 57-88; and: *Patent and Trademark Information: Uses and Perspectives* (ed: Virginia Baldwin) The Haworth Information Press, an imprint of The Haworth Press, Inc., 2001, pp. 57-88. Single or multiple copies of this article are available for a fee from The Haworth Document Delivery Service [1-800-HAWORTH, 9:00 a.m. - 5:00 p.m. (EST). E-mail address: docdelivery@haworthpress.com].

http://www.haworthpress.com/store/product.asp?sku=J122
© 2001 by The Haworth Press, Inc. All rights reserved.

KEYWORDS. *esp@cenet®*, European Patent Office, patents, full text

> ... [P]atents contain solutions to technical problems ... and represent an almost inexhaustible source of information.[1]

THE EUROPEAN PATENT OFFICE

The European Patent Office (EPO) (www.european-patent-office.org) is an international patent-granting authority established under the European Patent Convention (EPC) signed on October 5, 1973 in Munich, Germany, and entered into force on October 7, 1977.[2,3] The EPO is headquartered in Munich with a branch office in The Hague, The Netherlands, and sub-offices in Berlin, Germany and Vienna, Austria. The EPO is self-financed, with operating and investment budgets derived from processing fees and the annual renewal fees levied on granted European patents.[4] The EPO staff currently numbers 5,000 individuals and includes nationals of all contracting states.[5]

The European Patent Organisation, for which the European Patent Office acts as executive arm, currently has twenty-six (26) member states (see Table 1). These include all European Union (EU) countries as well the Czech Republic, Cyprus, Hungary, Liechtenstein, Monaco, Slovak Republic, the Republic of Bulgaria, the Republic of Estonia,

TABLE 1. List of Current Contracting Members of the European Patent Organisation and Their Associated Alphabetical Country Codes

Austria (AT)	Luxembourg (LU)
Belgium (BE)	Monaco (MC)
Cyprus (CY)	The Netherlands (NL)
Czech Republic (CZ)	Portugal (PT)
Denmark (DK)	Republic of Bulgaria (BG)
Finland (FI)	Republic of Estonia (EE)
France (FR)	Slovak Republic (SK)
Germany (DE)	Slovenia (SI)
Hellenic Republic (GR)	Spain (ES)
Hungary (HU)	Sweden (SE)
Ireland (IE)	Switzerland (CH)
Italy (IT)	Turkey (TR)
Liechtenstein (LI)	United Kingdom (GB)

Switzerland, and Slovenia and Hungary, the most recent members.[6,7] The protection conferred by European patent applications and granted patents has also been extended to a number of central and eastern European states; these presently include Albania (AL), Lithuania (LT), Latvia (LV), the former Yugoslav Republic of Macedonia (MK), and Romania (RO).[8]

The EPO grants European patents under a uniform and centralized procedure. In filing a single patent within the EPO framework, an applicant can obtain patent protection in as many EPO member and extension states as designated. A patent is normally granted for 20 years, although patents relating to pharmaceutical and plant protection products can be extended. Applications can be submitted in any of the three official languages of the EPO: English, French, or German.[9] The European Patent Convention is coupled with the Patent Cooperation Treaty (PCT), which also offers a uniform and simplified filing procedure in more than 100 countries.[10,11]

esp@cenet®

To improve access to its collections and services for the public and the non-specialist, the EPO in October 1998 formally inaugurated *esp@cenet®*, a free Web-based search and document service.[12,13] To utilize this service, users are required to use a JavaScript-enabled browser (Netscape Navigator 3, or higher; Microsoft Internet Explorer 3, or higher; or equivalent browsers). In addition, the Adobe® Acrobat® Reader® (version 3 or higher) must be installed to read 'facsimile images,' the full-text of associated patent documents in Portable Document Format (PDF).[14]

Within *esp@cenet®* (www.european-patent-office.org/espacenet/), a user can search available patent database collections using one of three gateways: via an EPO gateway; via the national patent offices of member states; or, via a European Commission (EC) gateway.

The first and third gateway options enable a user to search:

- a worldwide patent database;
- the patent database of the EPO;
- the patents issued by the World Intellectual Property Organization (WIPO); or
- a database of Japanese patents.

The second allows the user to search the national patent databases of EPO member states as well as the patent databases available in the first and third gateways. While the language support from the second gateway will vary from national office to national office, the first and third gateways offer trilingual interfaces (English, French ('Français'), and German ('Deutsch')). Within any of the three gateways, the user is offered a basic interface ('Quick Searches') (see Figure 1) and access to an advanced interface ('Worldwide–30 million documents') for searching a "worldwide" database collection (see Figure 2).[15]

EPO GATEWAY

Among available options, the EPO gateway (ep.espacenet.com) allows the user to perform 'Quick Searches' in the *esp@cenet®* *worldwide* database, a patent database that provides access to more than 42.5 million records from more than 70 countries, regions, or offices, includ-

FIGURE 1. Top Half of the European Patent Office (EPO) Gateway Homepage with Search Form for Performing 'Quick Searches' in the '*esp@cenet* ® Worldwide Patent Database'

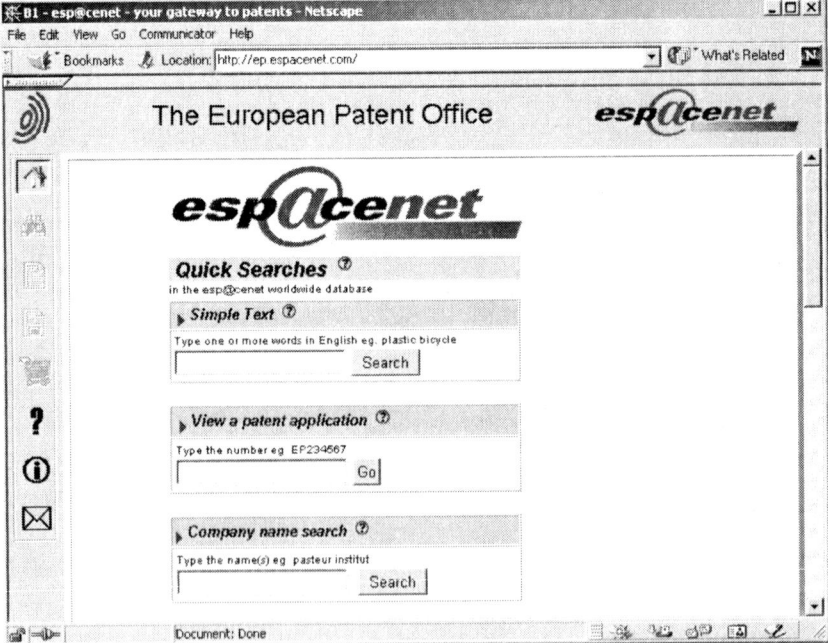

FIGURE 2. Bottom Half of the EPO Gateway Homepage Listing Other Accessible *esp@cenet* ® Patent Databases and a Patent Classification Search Tool ('ClassPat')

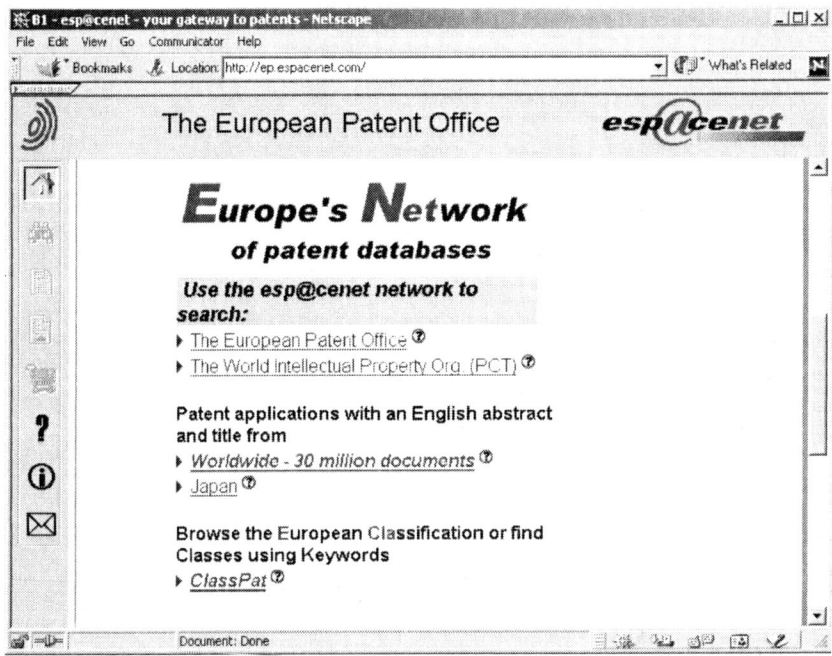

ing nearly 6.75 million entries for United States patent documents dating from the mid-19th century[16] and full text images for U.S. patents beginning in 1836.[17]

QUICK SEARCHES

Within the EPO gateway, three basic search alternatives are available using the Quick Searches option: 'Simple Text,' 'View a Patent Application' and 'Company Name Search' (see Figure 2).

'Simple Text'

The 'Simple Text' search allows the user to search the title and abstracts of incorporated patent documents, most notably patent applications; full-text document search capability is not currently available. Search terms and phrases must be entered in English. Users may input single search terms (e.g., 'computer'), search phrases, (e.g., 'pocket computer'), or simple Boolean (e.g., 'pocket AND computer') or nested Boolean statements (e.g., '(pocket OR palm OR handheld) AND computer'). A search phrase enclosed in double quotes (e.g., "pocket computer") will retrieve those records that have the phrase in the title and/or abstract of the record. A search of the phrase without double quotes will search for records that contain the first and secondary term(s) in the same record, regardless of their proximity. In *esp@cenet®*, the Boolean operator 'NOT' is available,[18] as is right truncation using established truncation symbols (i.e., '?', '#', '*') for selected fields (i.e., inventor, applicant, title, abstract).[19]

'View a Patent Application'

Within the Quick Searches option, users may also search for patent documents by patent document number (e.g., 'GB2301278'). The document number is composed of a two-letter country code (e.g., 'GB' for Great Britain) and the assigned number of the publication (e.g., '2301278'). In searching *esp@cenet®*, however, a patent number *kind code* may need to be included with the country code and document number. In the patenting process, a patent document may undergo a series of reviews, evaluations, approvals, and corrections, or may be of a particular type. A kind code is an alphabetic (e.g., 'A') or alphanumeric (e.g., 'B2') code that designates the stage of publication in the patenting process (for example, an unexamined patent application versus a published patent)[20] or a patent type (for example, a plant patent application), and is included in the full patent number. For example, in the patent document number 'DE 19932527 A1,' the country code ('DE') and kind code ('A1') collectively denote that this document is an unexamined German patent application. While in many countries the identical kind code may designate the identical type of patent document, the convention may not be uniform and users should consult appropriate documentation for the respective designation of each code.[21]

Initially, *esp@cenet®* primarily provided access to patent applications, and not granted patents.[22] In 2000, however, access to all publication levels for European (EP) was provided within the worldwide patent database, and now includes select granted patents.[23]

In searching *esp@cenet®* by patent document number, the 'A,' 'B,' and 'C' kind codes, as well as those alphanumeric codes that begin with any of these letters, are usually *omitted*; however for *certain* national patent documents, the 'B' and 'C' kind codes are *required*. Other types of kind codes *must be appended* to a patent document number to conduct a search for most–*but not all*–patent documents that contain these other types of kind codes. For example, to search for the German patent document numbered 'DE T 68926138,' the patent number must be entered as 'DE68926138T.' For German patent documents, the 'T' kind code indicates that the document is a translation of an international patent application.[24]

While generally excellent, the *esp@cenet*® online help system does not list kind codes nor provide assistance in their use. Although a downloadable manual includes information in searching *esp@cenet*® using kind codes, the documentation assumes knowledge and understanding of the codes and offers a limited number of examples as explanation for their inclusion or exclusion.[25]

'Company Name Search'

The 'Company Name Search' option allows the user to search for the patent documents associated with a particular company, corporation, university, or other named organization. As there is no authority control for corporate names and the form of an organizational name may vary, users should consider all possible forms when searching (e.g., '3M' or 'Minnesota Mining and Manufacturing'). The 'Company Name Search' option is a simplified alternative for searching the 'Applicant' field of the database document in the worldwide patent database and thus also allows users to search for a personal patent applicant name as well (e.g., 'Ali Selimoglu').[26]

SEARCH RESULTS

Upon execution, the results of a search in any of the Quick Searches options are displayed in the original search window ('B1') and include the following data and information (see Figure 3):[27]

- search criteria used (e.g., 'plastic AND bicycle');
- the field name(s) (e.g., 'TITLE OR ABS') [Title or Abstract];
- the number of documents found that match the search criteria (e.g., '327');

- the patent document numbers of the listed document records (e.g., 'GB699734,' 'WO0119668,' 'DE19932527,' etc.);
- the titles of the patent publication in English, where available (e.g., 'Advertising plaques for bicycles consists of trapeze-shaped plastic/metal element with folded edges for fastening to wheel spokes').

Results are listed in reverse order by date of inclusion in the original worldwide *esp@cenet®* database, not in chronological order or reverse chronological by 'publication' date. Search results are displayed in groups of twenty (20) entries and may be navigated using a 'jump bar' ('1 2 3 4 . . .') at the top of the results list (see Figure 3). Within *esp@cenet®*, search results are limited to a display of 500 entries. When there are no matching documents, an appropriate message appears in place of the results list.[28]

FIGURE 3. Search Results for a Sample 'Simple Text' Search

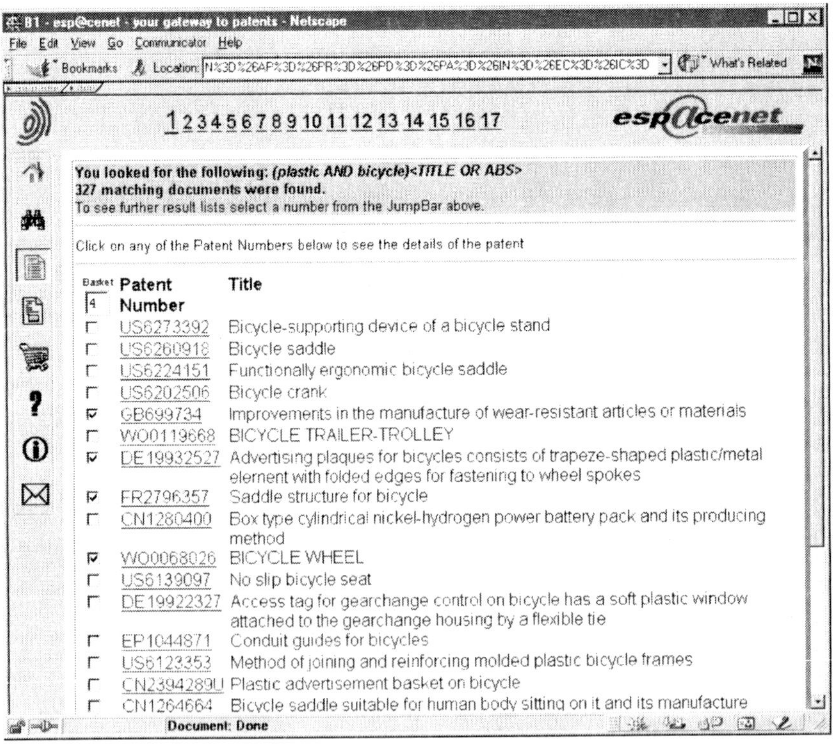

The bibliographical details and an English language abstract are displayed for a patent document by clicking a hyperlinked patent document number from the search results list (e.g., 'DE19932527').[29] Where available, the following types of bibliographic data are provided for the document in a separate browser window (B2) ('B2 *esp@cenet*–Document Viewer Navigation'):

- patent title in English (if available);
- patent number (e.g., 'DE19932527');
- publication date (e.g., '2001-02-01');
- inventor name(s) and nationality (e.g., 'Selimoglu Ali (DE)');
- applicant name(s) and nationality (e.g., 'Selimoglu Ali (DE)');
- requested patent (e.g., 'DE19932527');
- application number (e.g., 'DE19991032527 19990712');
- priority number(s) (e.g., 'DE19991032527 19990712');
- International Patent Classification (IPC) code(s) (e.g., 'B62J39/00'; 'B62J6/00');
- European Patent Classification (ECLA) code(s) (e.g., 'B62J39/00', 'B62J6/20');
- equivalents (if available);
- abstract in English (if available).

CLASSIFICATION

At EPO, two related, but different systems are used to characterize and classify the nature and content of patent documents: the International Patent Classification (IPC) scheme and 'European Patent Classification' (ECLA), the classification scheme developed by EPO. Under terms of the European Patent Convention, the EPO is required to apply the IPC classification (Rule 8),[30] a hierarchical classification system in which the field of technology is divided into approximately 70,000 categories.[31] If there are changes in a classification category in subsequent editions of the scheme, the IPC classifications previously assigned to a patent document are not revised by the EPO.[32] The IPC is administered and published by the World Intellectual Property Organization (WIPO) and is revised every five years. The current edition (7th) was issued on January 1, 2000.[33]

In addition to using IPC classification codes, the EPO assigns codes to patent documents from the European Patent Classification scheme (ECLA).[34] This classification contains 130,000 subdivisions, nearly twice as many as found within the IPC scheme.[35] Using the ECLA scheme, patent documents from around the world are classified within a particular technical field by the European patent examiner responsible for searches in that field, providing inherently more coherent and consistent classification designations. Unlike the IPC practice, a single version of the ECLA classification applies to the entire set of EPO patent documents that have been classified using the ECLA scheme. When changes occur in the scheme to reflect technological developments, the assigned classifications are revised to reflect these changes. While ECLA codes have been assigned to EPO incorporated patent documents dating from 1920, documents published prior to 1968 have not been classified using the IPC.[36] As of mid-October 2001, 16.5 million records in the worldwide patent database had received an ECLA classification.[37] Within *esp@cenet®*, patent document records with an ECLA code are hyperlinked from the code field to an associated section within a browsable copy of the classification.

In 2002, *ClassPat*, a feature that further facilitates use of ECLA, was introduced. ClassPat not only allows users to directly browse the ECLA scheme, but also provides a search function that retrieves a set of annotated codes, in rank order, that have the greatest association with given search terms or phrases.[38,39] The user can then subsequently conduct a search within the ClassPat interface of the worldwide patent database using classification codes of interest ('Search using selected Classifications').

PRIORITY

It is not uncommon for an inventor to seek to patent the same or similar invention in other countries or regions. In general, when an individual seeks to patent such inventions in more than one country, any application or patent published after the original application or patent is deemed *equivalent* to the original.[40] Within *esp@cenet®*, for two documents to be described as 'equivalents,' all their *priorities* must be the same.[41]

The term 'priority' (or 'priorities') relates to the priority system first established under Article 4 of the Paris Convention for the Protection of Industrial Property in 1883.[42,43] In general, the priority date is the date on which a patent application is filed or the filing date of another patent to which the subsequent patent claims the benefit of priority. Normally, the priority date is that given the applicant's domestic patent application and is often used in evaluating the 'novelty' of an invention.[44] In general, an invention is considered to be 'novel' if it is "... different in some way from all previous inventions."[45,46] Within this framework, an applicant is entitled to retain the priority of a first application filed in a country that is a signatory party to the Paris Convention for subsequent applications with respect to the same invention filed in other Convention countries within a period of twelve months from the date of the initial filing. Where an application validly claims the priority of a previous first application, it is the date of filing of that application ('priority date') that is the effective one for determining the state-of-the-art against which an invention's claims are evaluated, not the filing date of the subsequent application.

A *priority number* is the number of an earlier application whose priority has been claimed, and in the context of patent searching provides the common link between patent documents belonging to the same patent family.[47,48] The main method used to search for equivalents in other countries is by means of the patent family system. Broadly speaking, a patent family consists of all patent documents ('family members') linked by a common priority number.[49,50,51,52]

RESULTS DISPLAY AND FULL-TEXT ACCESS

From within the B2 browser window, the user can display, when available, a description of the patent item and its *claims*, in English, or the original language of the patent document (e.g., German); a key drawing, if appropriate; and a 'facsimile' (full-text) page from the patent document. A patent 'claim' "... describes the structure of an invention in precise and exact terms.... Patent claims serve as a way ... to determine whether an invention is patentable.... In concept, a patent claim marks the boundaries of the patent in the same way as the legal description in a deed specifies the boundaries of the property."[53]

For Quick Searches, the availability of bibliographic data and the full-text of patent documents as 'facsimile images' in Portable Document Format (PDF) varies with the country and agency (see Table 2 and Table 3).[54,55] For U.S. patents, *ep@cenet®* provides bibliographic records that date from 1859 and full-text facsimile images beginning in 1836.[56]

Within some search results, certain entries may not have a title ('No English title available'). The absence of an English title for an entry indicates that a non-English patent document has not yet been translated

TABLE 2. Bibliographic and Full-Text Coverage Data for Worldwide Documents for Select Contracting Countries

Country	Code	Bibliographic Data Start	PDF Start
Austria	AT	1925	Beginning
Belgium	BE	1923	1920
Cyprus	CY	1921	1998
Denmark	DK	1928	1920
France	FR	1902	1920
Germany	DE	1879	Beginning
Hellenic Republic	GR	1975	1976
Ireland	IE	1948	1929
Italy	IT	1933	1985
Luxembourg	LU	1945	1945
Monaco	MC	1957	Beginning
Netherlands	NL	1914	Beginning
Portugal	PT	1971	1980
Spain	ES	1968	1964
Switzerland	CH	1888	Beginning
Turkey	TR	1973	None
United Kingdom	GB	1863	1920

TABLE 3. Coverage Data for Worldwide Documents for Selected International Patent Agencies, Japan, and the United States

Agency or Country	Code	Bibliographic Data Start	PDF Start
European Patent Organisation	EP	1978	Beginning
World Intellectual Property Organization	WO	1978	1978
Japan	JP	1958	1920
United States	US	1859	1836

or that the document was published in a country that does not meet certain documentation requirements. Where an English translation is not available, those patents in French, German, or Spanish will make use of their source language title.[57]

Where available, the full-text PDF file for the selected patent document is retrieved by clicking the hyperlinked patent document number in the appropriate bibliographic record field ('Requested Patent') (e.g.,

'DE19932527'). The document itself is displayed in a third browser window (B3) (e.g., 'B3 esp@cenet–Facsimile image display DE19932527') (see Figure 4). From within the PDF page, the user may access any of the following specific sections of the original document by clicking the associated labeled navigation buttons (e.g., 'Biblio,' 'Desc,' 'Claims') found above the Adobe® Acrobat® Reader® navigation icons (see Figure 4):[58]

- bibliographic data (PDF page in the original source document on which bibliographic data is recorded) ['Biblio'];
- description (PDF of the first page of the description of the document) ['Desc'];
- claims (PDF of the first page of the claims of the documents) ['Claims'];
- drawing (PDF of the first page of the patent document figures or illustrations) ['Drawing'].

Users may also navigate document sections, or section parts, by using a supplemental drop-down navigation feature located between the Adobe® Acrobat® Reader® navigation icons and the labeled navigation buttons (see Figure 4). This navigation feature also indicates the total number of pages within a document and offers page-by-page navigation.[59] During the course of searching and retrieval, all primary and secondary browser windows remain open. By selecting the appropriate icon found within a left-side vertical navigation bar in a B1 (see Figure 3) or a B2 window, users can bring the document view window and its content (B2) to the foreground, or return to the results lists (B1) (see Figure 3).[60]

Other patent databases and access options available through the EPO gateway include (see Figure 2):

- 'The European Patent Office' (patent applications for the most recent twenty-four (24) months from the EPO);
- 'The World Intellectual Property Org. (PCT)' (patent applications published for the most recent twenty-four (24) months published by the World Intellectual Property Organization (WIPO));
- 'Worldwide–30 million documents' (an advanced interface for searching the worldwide database); and
- 'Japan' (a database of Japanese patent publications).

FIGURE 4. The First Page of a German Patent Document (DE 199 32 527 A 1) in PDF Format, with Document Navigation Features Located Above the Standard Adobe® Acrobat® Reader® Navigation Icons

EUROPEAN PATENT OFFICE SEARCH

In addition to the Quick Searches options within the EPO gateway, users can also search records for patent documents issued by the European Patent Office for the current *twenty-four* (24) *month* period ('The European Patent Office') (see Figure 2). Documents assigned issued by the EPO are prefixed with the code 'EP' (e.g., EP1125836). Under normal circumstances, new documents are added to this database weekly. As of February 8, 2003, there were 154,931 documents in the EPO database, with the oldest (EP1073600) published on February 7, 2001 and the most recent (EP1282350) published on February 5, 2003.

The 'European Patent Office' database is searched using a form that offers a variety of data fields, and these include (see Figure 5):[61,62]

- title (e.g., 'bicycle gear');
- publication number (e.g., 'EP0943536');
- application number (e.g., 'EP19990105514');
- priority number (e.g., 'DE19881011491');
- publication date (e.g., '1990922');
- applicant (e.g., 'ICI');
- inventor (e.g., 'Wang');
- IPC classification (e.g., 'B62M1/04').

For most fields, if more than one term is entered, the terms are combined in an implied 'AND' Boolean relationship. In the case of the

FIGURE 5. Form for Searching the European Patent Office (EPO) Patent Documents Database

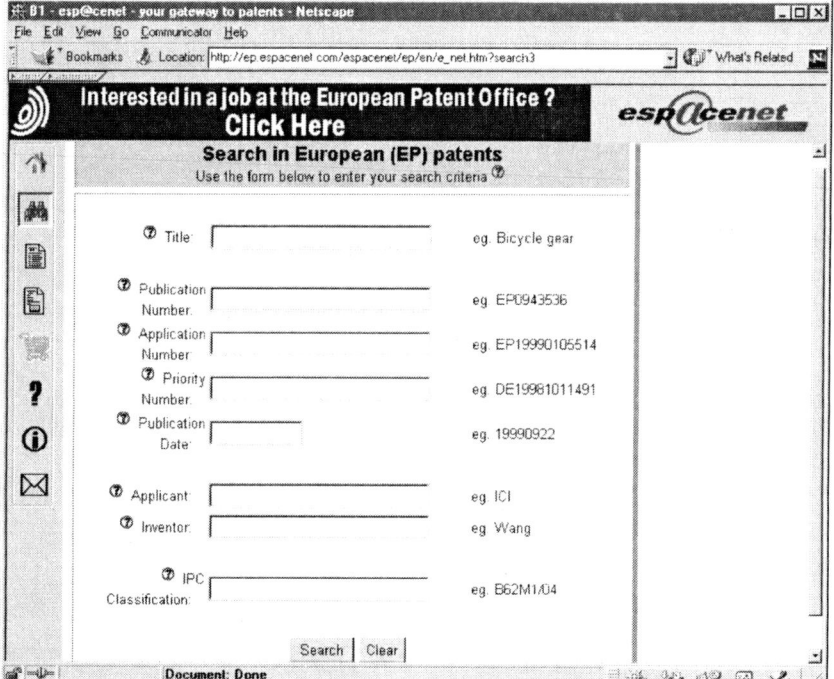

'Publication Number' and 'Application Number' fields, however, an 'OR' Boolean operator is applied. The explicit use of Boolean operators in any field, however, will override the default Boolean operator for the field. If criteria are entered in more than one input field, these will be combined in an 'AND' Boolean relationship. For the date field, only a single date (e.g., day, month, year) can be entered. To conduct a title search, the text must be entered as English language terms or phrases.

As with a Quick Searches option, results are listed in the original search window (B1) after execution. There is no discernable order, however, to the results listing; entries are neither listed in order by date of publication, alphabetically by the English title of the patent document, nor by inventor or applicant. And, as in a Quick Searches option, the bibliographical details for a patent document are displayed by clicking a hyperlinked patent document number from the search results list. Where available, the following bibliographic data are provided for the document in a separate browser window (B2) that is labeled 'B2 esp@cenet–Document Viewer Navigation':

- title (e.g., 'Gear device and method for driving a bicycle');
- requested patent (e.g., 'EP1183178');
- publication date (e.g., '2002-03-06');
- inventor(s) and nationality (e.g., 'Tushuizen Anthonius Johannus B (NL)');
- applicant(s) and nationality or country (e.g., 'Tushuizen Anthonius Johannus B (NL)');
- application number (e.g., 'EP20000927966 20000510');
- priority number(s) (e.g., 'WO2000NL00305 20000510; NL19991012008 20000510'); and
- IPC classification (e.g., 'B62M9/14').

In addition, a published WIPO document number may be included in the record (e.g., 'WO0068068') for some EP patents. This occurs when an EPO application is filed as part of a Patent Convention Treaty (PCT) patent application process (see next section).[63]

From within the secondary browser window (B2), the user can display a description of the patent item in its original language, and retrieve identical document components (e.g., claims, search report, key drawings, etc.) using the identical navigation options as made available in the Quick Searches option.

THE WORLD INTELLECTUAL PROPERTY ORGANIZATION (PCT) SEARCH

Within the EPO gateway, users can also search records for patent documents from the World Intellectual Property Organization (WIPO), an agency of the United Nations established to protect the intellectual property rights of creators and owners of its member states. Currently, 179 nations–more than 90 percent of countries of the world–are member states of WIPO, and include all member states of the European Patent Organization.[64] Among its designated responsibilities, WIPO administers international intellectual property treaties, including the Patent Cooperation Treaty (PCT).[65]

The PCT makes it possible for a national or resident of a PCT member state to seek patent protection for an invention simultaneously in each of the member states by filing an 'international' patent application. Under the PCT, an inventor is able to file a single international patent application that is legally recognized in the countries bound by the treaty and designated by the applicant. The effect of such an application in each designated state is the same as if a national patent application had been filed with the appropriate national office, thus providing a cost-effective method for filing and patent searching across many national jurisdictions. The overall intent, goal, and benefit of protection under the PCT is identical to that sought through the European Patent Office, but significantly broader in geographic and jurisdictional coverage. As of January 15, 2003, there were 118 contracting states.[66] Over a twenty-three year period, the number of international applications filed annually has increased from 2,600 in 1979 to more than 100,000 in 2001.[67,68]

Within *esp@cenet®*, WIPO patent documents for the most recent *twenty-four* (24) *month* period can be searched using a form with fields identical to those used to search the EPO database. As with the EPO search, titles must be searched using English language terms or phrases. Under normal circumstances, new documents are added to the WIPO database weekly, two weeks after formal publication. As of February 8, 2003, there were 139,387 documents in the *esp@cenet®* WIPO database, with the oldest (WO0078652) published on January 29, 2001 and the most recent (WO02104059) published on December 27, 2002. For earlier WIPO documents, users should search the *esp@cenet®* worldwide patents database. WIPO itself offers the 'PCT Electronic Gazette,' a database that provides bibliographic information, abstract, and drawing(s) for PCT applications published since January 1, 1997 [69] as well as a 'PCT Full Text Database Prototype,' which is planned to supersede the Gazette database after the full-text alternative is fully tested and evaluated.[70]

As with an EPO database search, results from a 'World Intellectual Property Org. (PCT)' search are listed in the original search window (B1) after execution. Once again, however, there is no discernable order to the results list. The bibliographical details for a patent document are retrieved by clicking a hyperlinked patent document number (e.g., 'WO0134456') from the list. The record data is displayed in a separate browser window (B2) ('B2 *esp@cenet*–Document Viewer Navigation') and is nearly identical in type as provided for an EPO patent document. From within this secondary browser window, users have the identical retrieval and navigation options made available from an EPO search.

'WORLDWIDE–30 MILLION DOCUMENTS' PATENT SEARCH

For users who wish to search the worldwide patents database using specific search fields not offered in the Quick Searches alternative, the EPO gateway offers a search form with numerous options (see Figure 6).[71] In addition to the search fields available in the EPO and the WIPO search forms, this worldwide patents database search form provides fields that allow users to search a title and an abstract concurrently and to search the database using a European patent ECLA classification code. As with other *esp@cenet*® patent databases, users are required to enter title and abstract terms and phrases in English. The worldwide patents database is updated weekly.[72]

As with other *esp@cenet*® database search results, entries from this worldwide patents database search are listed in the original search window (B1) after execution, in reverse order by date of inclusion in the original *esp@cenet*® worldwide database. And, as with other searches, the bibliographical details for a patent document are displayed by clicking a hyperlinked patent document number from the search results list. The bibliographic data types provided for a document are identical to those displayed in the results of a Quick Searches and likewise displayed in a separate browser window (B2) ('B2 esp@cenet–Document Viewer Navigation'). As with the Quick Searches option, the user can display, retrieve, and navigate the bibliographic record for a document and its segments, as well as the full-text of the document in PDF, within the B2 browser window.

FIGURE 6. Form Used for Searching the *esp@cenet* ® Worldwide Patent Documents Database

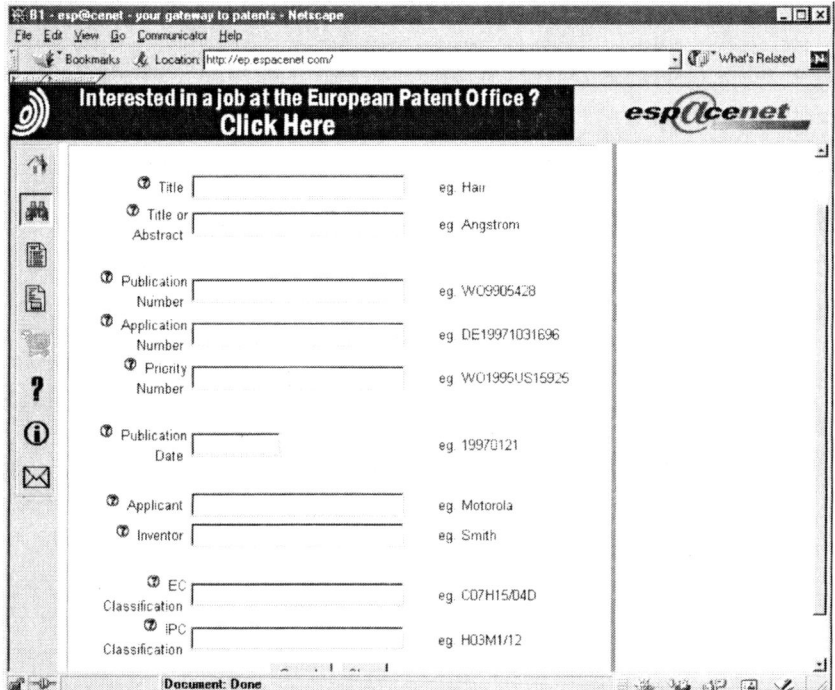

JAPAN (JAPANESE PATENT DOCUMENT SEARCH)

Among the most notable features of *esp@cenet*® is the separate access provided to Japanese patent documents.[73] Using a form nearly identical to that used to search the worldwide patent database, users can search the titles and abstracts of Japanese patent documents in English; by IPC code; and other standard options (e.g., publication number, application number, applicant, etc.). Patent abstracts date from October 1976, and first page document images, where available, begin in 1980.[74] While it is updated monthly, access to current Japanese patent publications is postponed for six months after official publication due to a translation delay.

As with other *esp@cenet*® database searches, results from a search in the Japanese patent database are listed in the original search window

(B1) after execution, with entries displayed in reverse alphanumeric order by the patent number. Entries are displayed in groups of twenty (20) and can be navigated, as with other database search results, using a jump bar at the top of the results list. Search results, as in other *esp@cenet®* databases, are limited to a display maximum of 500 entries.

As in Quick searches, the bibliographical details for a patent document are displayed by clicking a hyperlinked patent document number (e.g., 'JP2001170375') from the search results list. Where available, the following data types are provided for the document in a separate browser window (B2) ('B2 *esp@cenet*–Document Viewer Navigation'):

- patent number (e.g., 'JP2001170375');
- publication date (e.g., '2001-06-26');
- inventor(s) (e.g., 'Ogawa Hitoshi');
- applicant(s) (e.g., 'Matsushita Electric Works Ltd.');
- requested patent (e.g., 'JP2001170375');
- application number: (e.g., 'JP19990356555 19991215');
- priority number(s);
- IPC Classification (e.g., 'B26B19/44');
- EC Classification;
- Equivalents;
- abstract, in English.

As with other *esp@cenet®* record displays in the B2 browser windows, labeled navigation buttons are available and located in the upper left-hand corner of the page. However, unlike the range of navigation and access options found in other displays, records in the secondary browser window in the Japanese patent database generally have only one labeled navigation button ('Biblio'). For select records, however, a second labeled navigation button ('Page 1') may also be available. The 'Biblio' button redisplays the B2 page with its bibliographic data, while an adjacent second button ('Page 1'), when available, displays a sample page from the original patent document in PDF format.

NATIONAL PATENT OFFICE SITES

The second major gateway within *esp@cenet®* is that which provides direct or indirect access to the individual national patent databases and services offered by member and invited states[75,76] (see Table 4).

TABLE 4. Web Addresses and Language Support Provided by European Patent Organisation Members and Invited Countries

Country	Web Address	Language Supported
Austria	http://at.espacenet.com	German
Belgium	http://be.espacenet.com	French, Dutch
Bulgaria	http://bg.espacenet.com	Bulgarian
Cyprus	http://cy.espacenet.com	English
Czech Republic	http://cz.espacenet.com	Czech
Denmark	http://dk.espacenet.com	Danish
Finland	http://fi.espacenet.com	Finnish
Estonia	http://ee.epacenet.com	Estonian
France	http://fr.espacenet.com	French
Germany	http://de.espacenet.com	German
Hellenic Republic	http://gr.espacenet.com	Greek
Hungary	http://hu.espacenet.com	Hungarian
Ireland	http://ie.espacenet.com	English
Italy	http://it.espacenet.com	Italian
Latvia	http://lv.espacenet.com	Latvian
Liechtenstein	http://li.espacenet.com	French, German, Italian
Lithuania	http://lt.espacenet.com	Lithuanian
Luxembourg	http://lu.espacenet.com	French
Monaco	http://mc.epsacenet.com	French
Netherlands	http://nl.espacenet.com	Dutch
Poland	http://pl.espacenet.com	Polish
Portugal	http://pt.espacenet.com	Portuguese
Romania	http://ro.espacenet.com	Romanian
Slovakia	http://sk.espacenet.com	Slovakian
Slovenia	http://si.espacenet.com	Slovenian
Spain	http://es.espacenet.com	Spanish
Sweden	http://se.espacenet.com	Swedish
Switzerland	http://ch.espacenet.com	French, German, Italian
Turkey	http://tr.espacenet.com	Turkish
United Kingdom	http://gb.espacenet.com	English

From within the *esp@cenet®* homepage of a member state (e.g., United Kingdom), the user is offered access to databases identical to those offered within EPO gateway (see Figure 7). In addition, access is provided, in general, to the national patent database for a selected member state (e.g., 'Great Britain') and, indirectly, to the individual patent databases of other select European member states ('Other European countries') (see Figure 8). The national databases provide access to the bibliographic records for the current *twenty-four* (24) *months* of their respective national patent publications along with their associated full-text in PDF format, when available. Under normal circumstances, new

FIGURE 7. Top Portion of the *esp@cenet* ® Homepage for the National Patent Office of the United Kingdom (The Patent Office)

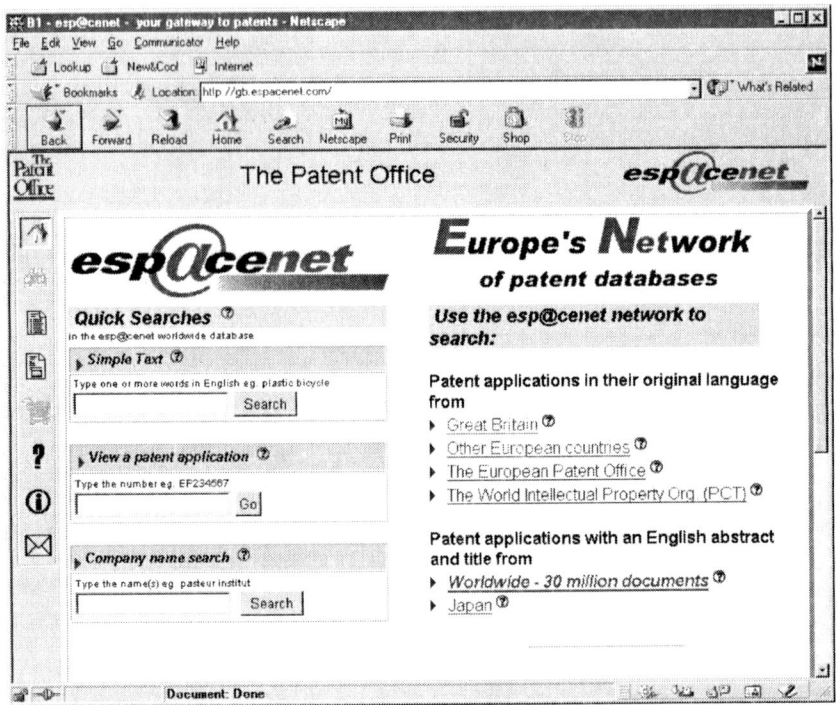

documents are added to a national database weekly, on the official day of publication of the paper document. For national documents older than 24 months, users should search the worldwide patents database.[77]

Although the language of the search form will vary by member state (see Table 4), the available search fields within the form are identical to that used for searching the European Patent Office patent database (see Figure 5). As with a search in the EPO database, results are listed in the original search window (B1) after execution and the bibliographical details for a patent document are displayed by clicking a hyperlinked patent document number (e.g., 'GB2354495') from the search results list. The data types for a document are identical to that provided by an EPO search and displayed in a separate browser window (B2) ('B2 *esp@cenet–Document Viewer Navigation*').

FIGURE 8. Form for Searching the National Patent Databases of 'Other European Countries' Showing a Drop-Down Menu of Available National Web Sites

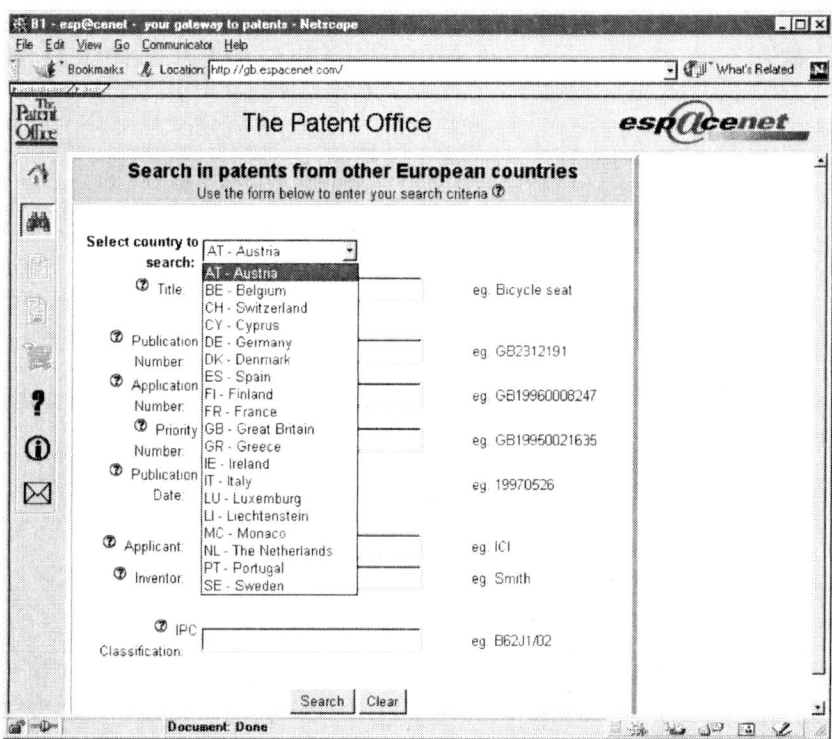

From within this secondary browser window (B2), the user can display the bibliographic data for the document ('Biblio') (a redisplay of the original B2 record) and retrieve the first page of the document; a description of the patent item in its original language; its claims; a key drawing, if appropriate; and search report ('S.R.'), if available, in PDF. As with EPO results, users can also navigate document sections, or section parts, by using the supplemental drop-down navigation feature.

RECENT AND FUTURE DEVELOPMENTS

Since its public introduction, *esp@cenet®* has been continuously refined and enhanced. In October 1999, a new version introduced several additional functionalities, including:[78]

- the automatic transfer of search criteria to subsequent forms;
- an enhanced online 'help' system that includes up-to-date content;
- the availability of search and display fields for the European patent classification number (ECLA); and
- the opportunity to mark a search result set and to add these to a 'shopping basket.'

The latter functionality is the foundation of a planned fee-based document delivery service that will complement the current free page-by-page facsimile delivery feature.[79] The contents of the shopping basket will form the basis of an 'order' and the full-text of selected documents, along with their associated figures, will be delivered in paper, by mail or electronically from an online download.[80]

In October 2001, *epoline®*–"a secure and integrated environment for electronic communication between the EPO and the applicants, their representatives and the national patent offices of the European Patent Convention (EPC) contracting states"–was formally launched.[81,82] *Epoline®* (www.epoline.org) provides access to the *European Patent Register*, as well as to an online patent application filing system, file inspection database, and several related services.[83] The patent register is a database containing all published European patent applications and published international applications (PCT) applying for a European patent in one or more member states of the European Patent Convention.[84,85] From within its form the following fields can be searched: publication number, application number; priority number; applicant, inventor, representative, and/or opponent name(s); an IPC classification; and associated publication, application, and priority number dates (see Figure 9).[86]

More recently, in Spring 2002, a 'Cited Documents' functionality was introduced within EP and PCT *esp@cenet®* application records. By clicking on the hyperlinked cited document identifiers, users can easily access those patent documents that a patent examiner has determined are highly related to the application, thereby facilitating ready access to associated technologies.[87]

ONLINE HELP SYSTEM

One of the primary aims of *esp@cenet®* is to provide open and free access to patent documentation and collections for the public and non-

FIGURE 9. Form Used to Search the *Online European Patent Register*

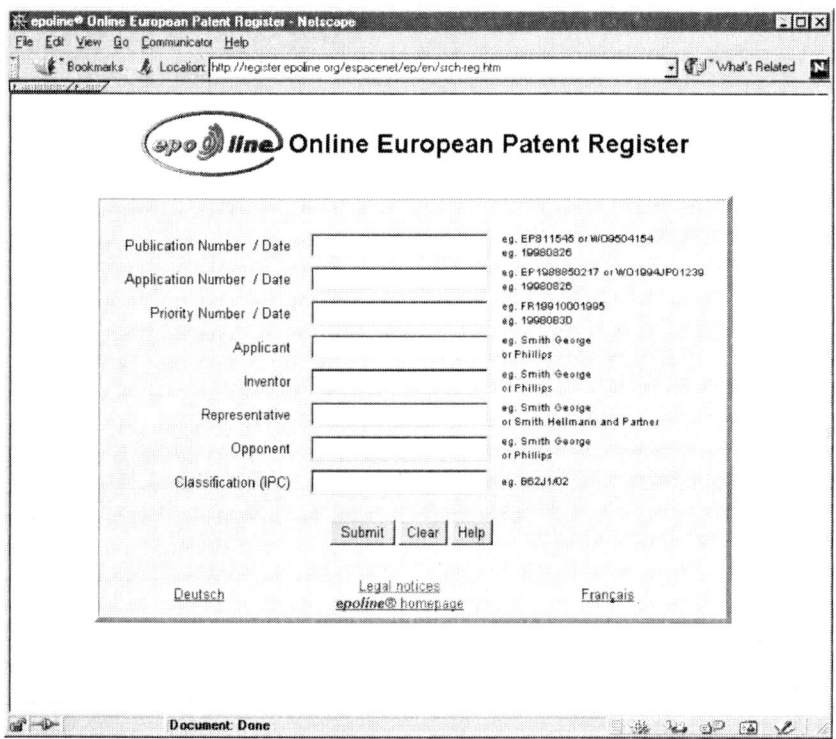

specialists. To facilitate use of these resources, *esp@cenet®* users are provided with an outstanding online 'help' system that offers in-context as well as general assistance (see Figure 10). In general, the text is well written, the layout well designed for easy scanning and review, and sufficient examples are provided to illustrate a search requirement or technique. Within a section, selected text is hyperlinked to relevant sections of the help system or to specific subsections for a general topic (see Figure 11).

COVERAGE OF UNITED STATES PATENT DOCUMENTS

As previously noted, *esp@cenet®* provides access to nearly 6.75 million entries for United States patent documents dating from the mid-19th century, and full-text access to U.S. patents in PDF format begin-

FIGURE 10. Screen Print of the Table of Contents for the *esp@cenet*® Online 'Help' System

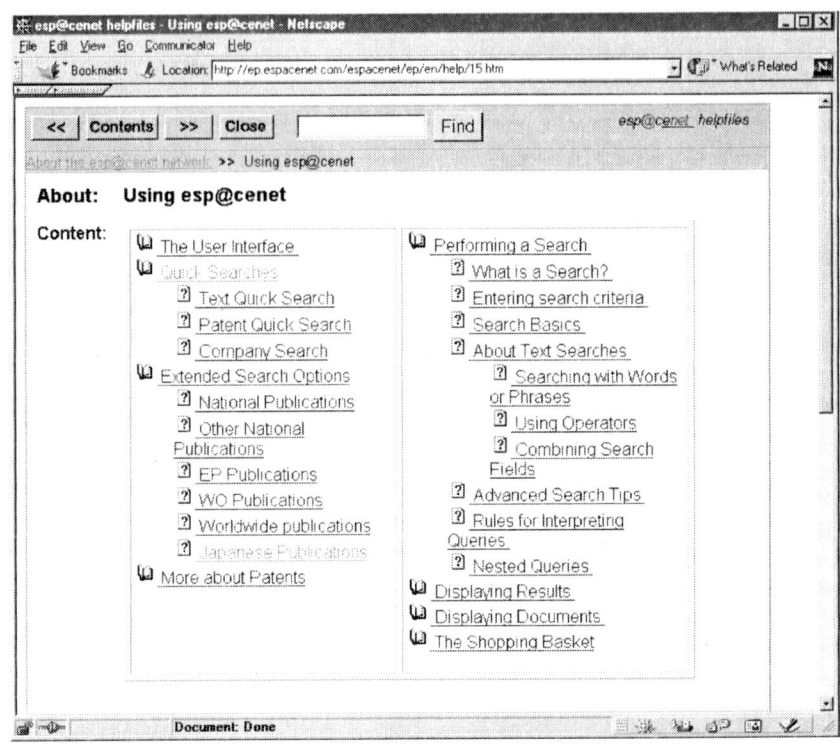

ning in 1836. In comparison, the United States Patent and Trademark Office (USPTO) (uspto.gov) offers full-text ASCII access to U.S. patent documents from 1976 and full-page images dating from 1790.[88] While the coverage of U.S. patents in *esp@cenet*® is significantly retrospective, it should not be viewed as a substitute for the complete coverage and the 'guidance, tools, and manuals' and other features and functionalities accessible from the USPTO site.[89]

INTEGRATED FRAMEWORK

In recent years, a variety of approaches have emerged that seek to facilitate use of Web-based resources. Among the more recent technologies are portals, unified interfaces that offer standard search and re-

FIGURE 11. Screen Print of Selected Text from a 'Help' File in *esp@cenet*®

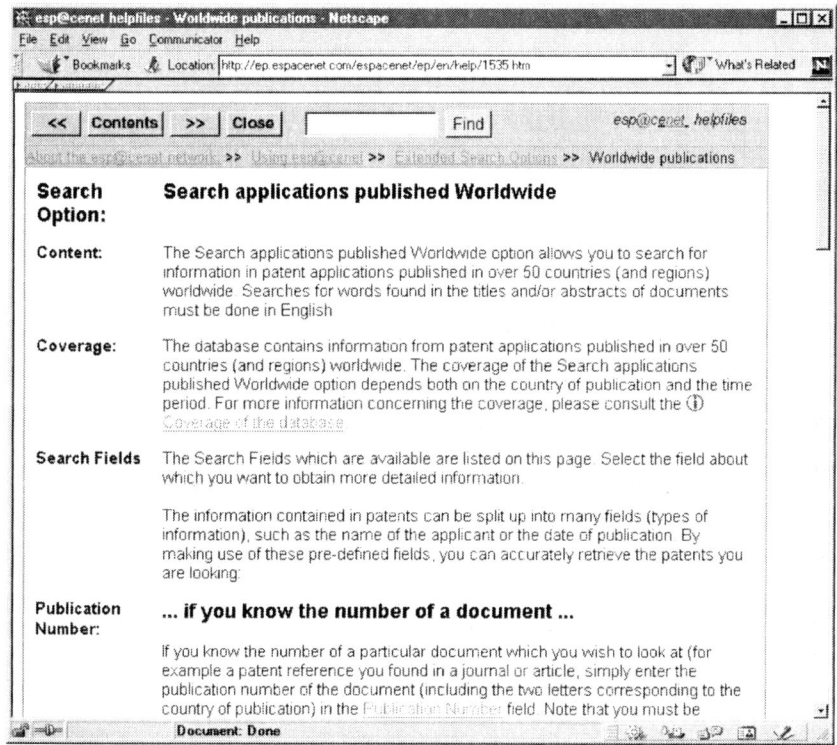

trieval options for accessing related or disparate digital resources. Noteworthy examples of portal technologies include *ENCompass* (Endeavor) (encompass.endinfosys.com), *MuseSearch* (MuseGlobal) (www.museglobal.com/Libraries/index.html), and *WebFeat* (WebFeat) (www.webfeat.org),[90] as well as *MetaLib* (ExLibris) (aleph.co.il/metalib/index.html). A system that integrates related resources within a common configuration is yet another approach for enhancing resource discovery. *esp@cenet*® is a premier example of the latter approach.

Within a centralized framework, *esp@cenet*® users are provided access to more than 42.5 million bibliographic patent document records from more than 70 countries. In addition, a significant percentage of these records are linked to their associated full text, with coverage dating from the first national patent document in most cases. Alternatively, users can access a collection of regional (EPO) or international (WIPO) patents; search a collection of national patent documents in the official language(s) of one of several major European nations; or retrieve English-language abstracts for non-English patent documents (e.g., Japanese). Overall, *esp@cenet®* enables users to access a wide variety of patent documents utilizing identical or similar search and display interfaces. And, perhaps most significantly, *esp@cenet®* allows users to access comprehensive patent collections from their desktop without direct cost.

DISCLAIMER

While *esp@cenet®* is comprehensive in many respects, the information furnished within it is not exhaustive. Although *esp@cenet®* offers a wide variety of collections and services, it should not be viewed, however, as a substitute for a formal patent search conducted by trained patent professionals.

> Although great care is taken in the proper and correct provision of services, the *esp@cenet®* network does not accept any responsibility for possible consequences of errors or omissions in the provided information. Readers are advised to seek independent professional advice before acting on any of the provided information.
>
> The *esp@cenet®* network reserves the right to amend, extend or withdraw any part or all of the offered services without notice.[91]

REFERENCES

1. European Patent Office, "Introduction," c2002. <http://www.european-patent-office.org/espacenet/info/introduction.htm> (1 February 2003).
2. European Patent Office, "The European Patent Office," c2001. <http://www.europeanpatentoffice.org/epo/pubs/brochure/general/e/epo_general.htm> (2 February 2003).
3. Ladas & Parry, "Protecting Inventions in Europe," c1999. <http://www.ladas.com/Patents/epo.html> (1 February 2003).
4. European Patent Office, "Detailed Information about the EPO," 2001. <http://www.european-patent-office.org/epo_detailed.htm> (1 February 2003).
5. European Patent Office, "Facts and Figures 2001," 2002, 3 <http://www.european-patent-office.org/epo/facts_figures/facts2001/pdf/facts_figures_01.pdf> (1 February 2003).
6. European Patent Office, "EPO Member States," 2003. <http://www.european-patent-office.org/epo/members.htm> (1 February 2003).

7. European Patent Office, "Press Release: Slovenia and Hungary Join the European Patent Organization," 2002. <http://www.european-patent-office.org/news/pressrel/2002_12_02_e.htm> (1 February 2003).
8. European Patent Office, "EPO Member States."
9. European Patent Office, "Detailed Information about the EPO."
10. Ibid.
11. Ian Muir, Matthias Brandi-Dohrn, and Stephan Gruber, "The Paris Convention, the EPC and the PCT." In *European Patent Law: Law and Procedure under the EPC and PCT* (Oxford: Oxford University Press, 1999), 1-17.
12. European Patent Office, "*esp@cenet®*–Past, Present and Future," *EPIDOS News Online* 4/1999 (1999). <http://www.european-patent-office.org/news/epidosnews/source/epd_4_99/6_4_99_e htm> (1 February 2003).
13. European Patent Office, "Press Release: Patent Information on Internet," 1998. <http://www.european-patent-office.org/news/pressrel/dips1.htm> (1 February 2003).
14. European Patent Office, "*esp@cenet®*. Helpfiles: Web Browsers and Additional Software," 2001. <http://ep.espacenet.com/espacenet/ep/en/help/13.htm> (1 February 2003).
15. European Patent Office, "*esp@acenet®*: A New Era in European Patent Information. Access," 2002. <http://www.european-patent-office.org/espacenet/info/access. htm> (1 February 2003).
16. European Patent Office, "*esp@cenet®*. Helpfiles: Worldwide Database–Detailed Coverage (Updated 03/02/2003)," 2003. <http://ep.espacenet.com/espacenet/ep/en/help/tab2.html> (18 February 2003).
17. European Patent Office, "*esp@cenet®*. Helpfiles: Worldwide Databases: Facsimile and Classification Coverage," n.d. <http://ep.espacenet.com/espacenet/ep/en/help/tab1.html> (9 February 2003).
18. European Patent Office, "*esp@acent®:* The EPO Guide to Patent Information on the Internet," n.d., 40. <http://epart.epo.org/dwl/espacenet_manual.pdf> (7 February 2003).
19. European Patent Office, "*esp@cenet®*. Helpfiles: Searching With Words or Phrases in Titles and Abstracts," 2001. <http://ep.espacenet.com/espacenet/ep/en/help/15441.htm> (1 February 2003).
20. Dialog Corporation Training Center, "Patents: Country and Kind Codes in Files 345 and 351," c2003. <http://training.dialog.com/quick/solutions/4924.html> (1 February 2003).
21. Delphion, "Help: Patent Kind Codes by Country," c2001. <http://www.delphion.com/ help/kindcodes> (1 February 2003).
22. European Patent Office, "The *esp@cenet®* Helpdesk's Five Most Frequently Asked Questions," *EPIDOS News Online* 3/99 (1999). <http://www.european-patent-office.org/news/epidosnews/source/epd_3_99/9_3_99_e.htm> (1 February 2003).
23. European Patent Office, "New Features in *esp@cenet®:* Access to All levels of European (EP) Applications in Facsimile Mode," *EPIDOS News Online* 3/00 (2000). <http://www.european-patent-office.org/news/epidosnews/source/epd_3_00/3_3_00_e. htm> (1 February 2003).
24. European Patent Office, "*esp@acenet®:* The EPO Guide to Patent Information on the Internet," n.d., 42.
25. Ibid.
26. Ibid., 44.
27. Ibid., 28.
28. Ibid.

29. Ibid.
30. Ibid., 45.
31. World Intellectual Property Organization, "International Patent Classification (IPC)," n.d. <http://www.wipo.int/classifications/en/> (8 February 2003).
32. European Patent Office, "Using the 'EC classification' Field on the *esp@cenet®* Service," *EPIDOS News Online* 1/00 (2000). <http://www.european-patent-office.org/news/epidosnews/source/epd_1_00/14_1_00_e.htm> (8 February 2003).
33. World Intellectual Property Organization, "IPC 7 English Version. Sections," n.d. <http://classifications.wipo.int/fulltext/new_ipc/ipc7/eindex.htm> (1 February 2003).
34. European Patent Office, "European Patent Classification," n.d. <http://l2.espacenet.com/espacenet/ecla/index/index.htm> (1 February 2003).
35. Gérard Giroud, "EPO Documentation Strategy: Facing the Challenges of the Patent System," Presentation at *EPIDOS Annual Conference 2000*, October 20, 2000, Vienna, Austria, Slide 24, 2000. <http://db1.european-patent-office.org/dwl/eac2000/01_giroud.pdf> (1 February 2003).
36. European Patent Office, "Using the 'EC classification' Field on the *esp@cenet®* Service."
37. European Patent Office, "*esp@cenet®* News. Some Statistics," *EPIDOS News Online* 4/01 (2001). <http://www.european-patent-office.org/news/epidosnews/source/epd_4_01/7_4_01_e htm> (8 February 2003).
38. European Patent Office, "*esp@cenet®*. Helpfiles: ClassPat: A New Way of Accessing Patent Classification," 2002. <http://l2.espacenet.com/espacenet/help/ecla/1.htm> (7 February 2003).
39. European Patent Office, "*esp@cenet®:* Two Important New Features," *EPIDOS News Online* 3/02 (2002). <http://www.european-patent-office.org/news/epidosnews/source/epd_3_02/3_3_02_e. htm> (8 February 2003).
40. British Library, "Glossary of Patent Terms," n.d. <http://www.bl.uk/services/information/patents/gloss.html> (8 February 2003).
41. European Patent Office, "Patent Families–An Attempt to Approach the Miracle (Part 2)," *EPIDOS News Online* 2/01 (2001). <http://www.european-patent-office.org/news/epidosnews/source/epd_2_01/2_2_01_e.htm> (1 February 2003).
42. World Intellectual Property Organization, "Paris Convention for the Protection of Industrial Property," n.d. <http://www.wipo.int/clea/docs/en/wo/wo020en.htm> (9 February 2003).
43. Ian Muir, Matthias Brandi-Dohrn, and Stephan Gruber, "The Paris Convention, the EPC and the PCT."
44. British Library, "Glossary of Patent Terms," n.d. <http://www.bl.uk/services/information/patents/gloss.html> (8 February 2003).
45. Nolo.com, "Everybody's Legal Dictionary: Novelty," c2001. <http://www.nolo.com/lawcenter/dictionary/dictionary_listing.cfm/alpha/P> (8 February 2003).
46. British Library, "Glossary of Patent Terms."
47. European Patent Office, "*esp@cenet®:* The EPO Guide to Patent Information on the Internet," n.d., 52.
48. Ian Muir, Matthias Brandi-Dohrn, and Stephan Gruber, "Priority." In *European Patent Law: Law and Procedure under the EPC and PCT* (Oxford: Oxford University Press, 1999), 18-28.
49. European Patent Office, "*esp@cenet®:* The EPO Guide to Patent Information on the Internet," n.d., 52.

50. European Patent Office, "Patent Families–An Attempt to Approach the Miracle," *EPIDOS News Online* 4/00 (2000). <http://www.european-patent-office.org/news/epidosnews/source/epd_4_00/14_4_00_e.htm> (1 February 2003).

51. European Patent Office, "Patent Families–An Attempt to Approach the Miracle (part 2)."

52. European Patent Office, "Patent Family Systems," *EPIDOS News Online* 4/00 (2000) <http://www.european-patent-office.org/news/epidosnews/source/epd_4_00/4_4_00_e.htm> (1 February 2003).

53. Nolo.com, "Everybody's Legal Dictionary: Patent Claim," <http://www.nolo.com/lawcenter/dictionary/dictionary_listing.cfm/alpha/P> (9 February 2003).

54. European Patent Office, "*esp@cenet®*. Helpfiles: Worldwide Database–Detailed Coverage (Updated 03/02/2003)."

55. European Patent Office, "*esp@cenet®*. Helpfiles: Worldwide Databases: Facsimile and Classification Coverage."

56. Ibid.

57. European Patent Office, "*ep@cenet®:* The EPO Guide to Patent Information on the Internet," 29.

58. Ibid., 32.

59. European Patent Office, "New Features in *esp@cenet®:* Navigation within the Facsimile Document," *EPIDOS News Online* 3/00 (2000). <http://www.european-patent-office.org/news/epidosnews/source/epd_3_00/3_3_00_.htm> (1 February 2003).

60. European Patent Office, "*ep@cenet®:* The EPO Guide to Patent Information on the Internet," 27.

61. Ibid [18]-21.

62. European Patent Office, "Search in European (EP) Patents," n.d. <http://ep.espacenet.com/espacenet/ep/en/e_net.htm?search3> (1 February 2003).

63. European Patent Office, "New Features in *esp@cenet®:* Links to WO Publication Numbers for Euro-PCT Applications," *EPIDOS News Online* 3/00 (2000). <http://www.european-patent-office.org/news/epidosnews/source/epd_3_00/3_3_00_e.htm> (1 February 2003).

64. World Intellectual Property Organization, "General Information about WIPO," 2001. <http://www.wipo.org/about-wipo/en/gib.htm> (9 February 2003).

65. World Intellectual Property Organization, *Patent Cooperation Treaty (PCT) Done at Washington on June 19, 1970, amended on September 28, 1979, and modified on February 3, 1984, and October 3, 2001 (as in force from April 1, 2002)* (Geneva: World Intellectual Property Organization, 2002) <http://www.wipo.int/pct/en/texts/articles/atoc.htm> (9 February 2003).

66. World Intellectual Property Organization, "Global Protection System Treaties. Patent Cooperation Treaty (PCT). Contracting Parties," 2003. <http://www.wipo.org/treaties/documents/english/pdf/m-pct.pdf> (9 February 2003).

67. C. P. Auger, ed. *Information Sources in Patents* (London; New York: Bowker-Saur, c1992), 59-64.

68. World Intellectual Property Organization, "Basic Facts about the Patent Cooperation Treaty," 2002, 11. <http://www.wipo.int/pct/en/basic_facts/basic_facts.pdf> (9 February 2003).

69. World Intellectual Property Organization, "Intellectual Property Digital Library. Search Page," 2003 <http://ipdl.wipo.int/en/search/search.shtml> (1 February 2003).

70. Ibid.

71. European Patent Office, "Search in Patents Throughout the World," n.d. <http://ep.espacenet.com/espacenet/ep/en/e_net.htm?search5> (1 February 2003).
72. European Patent Office, *"esp@cenet®.* Helpfiles: Search Applications Published Worldwide," 2001. <http://ep.espacenet.com/espacenet/ep/en/help/1535.htm> (9 February 2003).
73. European Patent Office, "Search in Japanese patents," n.d. <http://ep.espacenet.com/espacenet/ep/en/e_net.htm?search6> (9 February 2003).
74. European Patent Office, *"esp@cenet®.* Helpfiles: Search Japanese Patent Publications," 2001. <http://ep.espacenet.com/espacenet/ep/en/help/1536.htm> (1 February 2003).
75. European Patent Office, *"esp@cenet®:* A New Era in European Patent Information. Access." 2002.
76. European Patent Office, "Press Release: Slovenia and Hungary Join the European Patent Organization."
77. European Patent Office, *"esp@cenet®.* Helpfiles: Search 24 Months of National Patent Publications," 2001. <http://ep.espacenet.com/espacenet/ep/en/help/1531.htm> (1 February 2003).
78. European Patent Office, *"esp@cenet®*–Past, Present and Future."
79. Ibid.
80. Graham Day, *"esp@cenet®,"* Presentation at *EPIDOS Annual Conference 2000*, October 20, 2000, Vienna, Austria, Slides 10-18. <http://db1.european-patent-office.org/dwl/eac2000/02_grahamday.pdf> (1 February 2003).
81. European Patent Office, "The First *epoline®* User Day," *EPIDOS News Online* 4/2001 (2001). <http://www.european-patent-office.org/news/epidosnews/source/epd_4_01/6_4_01_e.htm> (16 February 2003).
82. European Patent Office, "What is *epoline®*?" 2003. <http://www.epoline.org/epoline/Epoline?language=EN&page=whatisepoline&b=NS> (16 February 2003).
83. Ibid.
84. European Patent Office, "What is the Online European Patent Register?" 2000. <http://www.epoline.org/webreg/help/en/what_is_the_online_european_patent_register.htm> (23 October 2001).
85. Madeleine Lanier, *"epoline®:* The European Patent Register and Online File Inspections," Presentation at *EPIDOS Annual Conference* 2000, October 20, 2000, Vienna, Austria <http://db1.european-patent-office.org/dwl/eac2000/03_epoline.pdf> (1 February 2003).
86. European Patent Office, "European Patent Register," 2001. <http://www.epoline.org/epoline/Epoline?language=EN&page=register&b=NS> (1 February 2003).
87. European Patent Office, *"esp@cenet®:* Two Important New Features."
88. United States Patent and Trademark Office, "Patent Full-Text and Full-Page Image Databases," 2003 <http://www.uspto.gov/patft/index.html> (25 February 2003).
89. United States Patent and Trademark Office, "Patents Guidance, Tools & Manuals," 2002 <http://www.uspto.gov/web/patents/guides.htm> (25 February 2003>.
90. Gerry McKiernan, "Library Database Advisors: Emerging Innovative Augmented Digital Library Services," *Library Hi Tech News* 19 no. 4 (May 2002): 27-33.
91. European Patent Office, *"esp@cenet®.* Helpfiles: Conditions of Use," 2001. <http://ep.espacenet.com/espacenet/ep/en/help/11.htm> (6 March 2003).

Regional Patent Systems: A Challenge for the International Searcher

Stephen R. Adams

SUMMARY. A number of international agreements exist which provide for a single regional patent office to grant patents on behalf of its member states. The most well-known is the European Patent Office, but similar organizations exist, in working or embryonic form, in most parts of the world. Procedures at the offices covering the former Soviet Union, Africa, the Middle East and South America are described. The Patent Cooperation Treaty is not a true regional patent system, but represents a very significant move towards greater international harmonization in patent granting procedures. *[Article copies available for a fee from The Haworth Document Delivery Service: 1-800-HAWORTH. E-mail address: <docdelivery@haworthpress.com> Website: <http://www.HaworthPress.com> © 2001 by The Haworth Press, Inc. All rights reserved.]*

KEYWORDS. Patents, patent granting, patent examination, international cooperation, Europe, Africa, South America, Middle East

INTRODUCTION

It is a fundamental truth of working with patent information that 'there is no such thing as a world patent.' If a searcher is seeking to es-

---

Stephen R. Adams is Managing Director, Magister Ltd., Crown House, 231 Kings Road, Reading, Berkshire, RG1 4LS, United Kingdom (E-mail: stevea@magister.co.uk).

[Haworth co-indexing entry note]: "Regional Patent Systems: A Challenge for the International Searcher." Adams, Stephen R. Co-published simultaneously in *Science & Technology Libraries* (The Haworth Information Press, an imprint of The Haworth Press, Inc.) Vol. 22, No. 1/2, 2001, pp. 89-99; and: *Patent and Trademark Information: Uses and Perspectives* (ed: Virginia Baldwin) The Haworth Information Press, an imprint of The Haworth Press, Inc., 2001, pp. 89-99. Single or multiple copies of this article are available for a fee from The Haworth Document Delivery Service [1-800-HAWORTH, 9:00 a.m. - 5:00 p.m. (EST). E-mail address: docdelivery@haworthpress.com].

http://www.haworthpress.com/store/product.asp?sku=J122
© 2001 by The Haworth Press, Inc. All rights reserved.
10.1300/J122v22n01_06

tablish whether an invention has been granted patent protection on a world-wide basis, there is no 'super-index' that they can consult which will give a definitive answer. Up until the mid-1970s, the tools for world-wide patent retrieval were highly fragmented. If you needed to locate patents from five, ten or fifty different countries, you needed to search in five, ten or fifty different places.

The world has moved on since then, and there are now many more tools to assist the job of searching for patents from many nations. However, the unwary patent searcher is still faced with a further obstacle–that of regional patent systems. These organisations exist, at least in embryonic form, in most corners of the world, and make the apparently simple question, "Is there a patent on this technology in country X?" a somewhat more complex affair.

The basic principle behind regional patent systems is that a single organization is mandated to handle some or all of the functions of a patent office, on behalf of each member state. In some cases, regionalisation extends only to the preliminary administrative procedures, leaving the actual patent-granting procedures in the hands of the members. In other systems, the patent-granting and even some of the post-grant opposition or litigation stages are handled by a central office. At the extreme, regional patent systems can result in the effective abolition of the national patent offices for each of the members. The result for the searcher is that it becomes necessary to check two sources when trying to establish whether patent rights exist in certain countries–firstly the national patent system and secondly any regional patent system to which that country may belong. It is no longer sufficient to follow the "one country, one source" rule. The patents granted by the regional office have an equal standing with national patents. For example, section 77(1) of the United Kingdom Patents Act 1977 is quite explicit that ". . . a European patent (UK) shall . . . be treated. . . . as if it were a patent under this Act granted in pursuance of an application made under this Act . . ."

Regional patent systems are not a totally modern phenomenon. In the United Kingdom, prior to 1852, there were separate systems granting patents in England (which included Wales), Scotland and Ireland. The true United Kingdom patent did not come into existence until a year after that great innovation showcase, the Great Exhibition of 1851. More recently, the four Scandinavian countries (Denmark, Norway, Sweden and Finland) tried to set up a Nordic patent system in the 1950s, but without success. The first operating regional patent system in the modern era is the European Patent Office (EPO), founded on a diplomatic Convention signed in 1973. The EPO started operation in 1978, and is now recognised as one of the most influential patent offices in the world, alongside the U.S. and Japanese offices.

REGIONAL PATENT OFFICES AND SYSTEMS–A SURVEY

The regional systems in the world at the time of writing break down into three broad classes:

- Already operational
- Nearly operational
- Early stages of planning.

These will be considered in turn.

Operational Systems

The European Patent Office

The most significant regional patent office is the European Patent Office (EPO). At the time of writing, there are twenty-four full members and six associate members (so-called 'extension states') of the EPO:

- Full members: Austria, Belgium, Bulgaria, Cyprus, the Czech Republic, Denmark, Estonia, Finland, France, Germany, Greece, Ireland, Italy, Liechtenstein, Luxembourg, Monaco, the Netherlands, Portugal, Slovakia, Spain, Sweden, Switzerland, Turkey and the United Kingdom.
- Extension states: Albania, Latvia, Lithuania, Macedonia, Romania and Slovenia.

Four of the extension states (Latvia, Lithuania, Romania and Slovenia) are candidates to become full members during 2002-3, along with Hungary and Poland, which would bring the membership to a total of thirty plus two remaining extension states.

It should be noted that the EPO is operationally independent of the European Union institutions, and its membership differs quite markedly from the EU; at the time of writing, the EU has only fifteen members. There are links between the two systems, in that a commitment by a country to join the EU brings with it a requirement that the state will in due course join the EPO, but the reverse is not the case. In fact, for most of the last thirty years, the membership of the EPO has been larger than the EU, with the most industrially significant country being Switzerland.

The operations of the EPO are centred in Munich and Berlin (Germany), The Hague (Holland) and Vienna (Austria). Most of the substantive examination and all of the legal division (for appeals, etc.) is based in Munich, although The Hague is becoming more involved in this sphere as well. The office in The Hague has developed out of an international search agency, the Institute International des Brevets, which was already operating as a centralised novelty search centre for several northern European countries before the EPO was founded, and has an unparalleled collection of world patent literature.

In common with many other countries, the EPO operates a two-stage publication and examination procedure, generally referred to as "deferred examination." Under this system, each step is associated with a separate published document.

During the first stage, the application is filed at the EPO for a small fee. In most cases, these applications will be based upon an earlier filing, usually at the inventor's home (national) patent office within the previous twelve months, using the Paris Convention procedure. At this point, the EPO only checks to ensure that the contents comply with certain formal requirements (such as identifying at least one inventor) and allocates an application number. The search division will then arrange for a search of the literature, in order to identify the most closely-related known references. Approximately eighteen months after first filing, the application is published in its entirety, exactly as filed. It is allocated a publication number, preceded by the two-letter code "EP" to indicate that the EPO is the publishing authority, and may appear in English, French or German. If the search report is published simultaneously with the specification, the publication number is given an "A1" suffix. In some 40% of cases, the search report is not completed by the time the specification is ready to publish, so the main text is produced first, designated an "A2," and the search report is published independently at a later date, using the code "A3."

At this point, the applicant can decide to withdraw from any further processing. Typically, they may do so if the search report indicates that the subject matter is already well known and therefore their application is unlikely to be granted.

If however, the applicant wishes to proceed, they pay a further fee to request full or "substantive" examination. The subject matter is then processed in detail by one or more patent examiners, a process which can take anything two years upwards–its length varies greatly depending upon the complexity of the case.

If the application passes and is finally granted, the specification is published a second time, incorporating any amendments which may have been imposed during the examination procedure. The same publication number is used, but this time the granted patent is denoted by a "B1" suffix.

At this point, there is a 9-month window during which companies opposed to the grant of the patent have a right to appeal to the central authorities of the EPO to have the patent amended or revoked. If the case goes to opposition, it may be re-issued in amended form as a "B2" document.

Thus far, the EPO has replaced the functions of the national patent offices of its members. The granted patents issued by the EPO have an equal standing alongside the national patents issued by each state. When an applicant uses the EPO system, it is not mandatory for them to apply for patent protection in all member states. In terms of costs, it is generally reckoned that if an applicant wants protection in more than three of the members, it is more economical to use the EPO route. Hence, the individual national patent systems have become more used by smaller corporations wishing to obtain cover only in their home country or perhaps one additional foreign market. The decision by the applicant as to how many states they wish to be covered in an EPO application is called "designation," and a list of the so-called "designated states" appears on the front page of both EPO applications and the final granted patents. It is only really meaningful in the case of the grant, when the list of countries tells you definitively where protection is being sought.

After the 9-month opposition period, the EPO ceases to have any further legal interest in a granted patent, other than receipt of annuity fees to maintain the patent in force. If there is any legal action taken against a granted European patent outside of the common opposition period, it has to be done in each individual state where the patent is in force, and litigated in the national courts of each state. This is clearly an expensive business, and the outcome will depend to some extent upon national legal precedent and court procedures. A decision in one designated state (e.g., the United Kingdom) has no immediate impact upon the validity of the same European Patent in another state, such as Germany. Likewise, the patent may be granted an extension of term in one state but denied it in another. For this reason, it can be argued that the EPO is not a true supra-national patent system, but more akin to a bundle of national patents which happen to have been examined separately under one statute, and accepted by all the states concerned.

Eurasian Patent Office

This regional office (the EAPO) grew very quickly out of the confusion of the post-Soviet era in the early 1990s. Almost immediately after the formal disintegration of the Soviet Union on December 25, 1991, a number of the former autonomous republics, now independent states, expressed the desire to work together in the field of intellectual property. An initial body, the Interstate Council on the Protection of Industrial Property, was instrumental in the negotiations which led to the signing of the Eurasian Patent Convention in Moscow on September 9, 1994. It entered into force when the first three states (Belarus, Tajikistan and Turkmenistan) ratified it, followed shortly afterwards by the Russian Federation, Kazakhstan, Azerbaijan, Kyrgyzstan, Moldova and Armenia. The Office started operations on January 1, 1996, with headquarters in Moscow. Of the six former Soviet states which are not currently members, the three Baltic states (Lithuania, Latvia and Estonia) are more westward-looking, and as noted above are candidates to join the European Patent Office soon. The remaining three (Georgia, the Ukraine and Uzbekistan) may also choose to remain outside the system.

The operation of the EAPO is very similar to the EPO, using deferred examination with early publication at eighteen months, followed by substantive examination leading to a grant of patent with a term of twenty years from filing date. The EAPO has only a single official language–Russian–as opposed to the three used by the EPO, and its documents are distinguished by the publication prefix "EA." The first unexamined documents were issued in October 1996, and the first patent was granted in April 1997.

One significant difference from the EPO practice is that it is not possible to designate individual member states. All nine states are by default included in the coverage of any granted patent, and the only way of withdrawing protection is by subsequently not paying the renewal fees for any unnecessary countries.

OAPI

The abbreviation OAPI stands for the Organisation Africaine de la Propriété Intellectuelle (African Organisation for Intellectual Property), which is based in Yaoundé, in the Cameroon. It has seventeen member states drawn from the former French-speaking colonies of sub-Saharan Africa. It dates its origins from 1962, when the Malagasy Republic and eleven other states concluded the Libreville Agreement, to form the African and Malagasy Patent Rights Authority (OAMPI). Subsequently, the Malagasy Republic (Madagascar) withdrew and additional members joined a new arrangement, founded on the Bangui Agreement signed on March 2, 1977. The members at the time of writing are: Burkina Faso, Benin, Cameroon, Central African Republic, Chad, Congo, Côte d'Ivoire, Djibouti, Equatorial Guinea, Gabon, Guinea, Guinea-Bissau, Mali, Mauritania, Niger, Senegal, and Togo.

The Bangui Agreement entered into force on February 8, 1982, which marked the start of what is still a unique procedure in the world of intellectual property. Unlike the other regional systems discussed above, the Bangui Agreement envisages a single authority created by a group of sovereign states, to both grant *and* litigate in the patents arena. As the name of the Organisation suggests, the most recent revision of the Agreement also has effect in other areas of intellectual property, such as trademarks, industrial designs and plant variety rights. The terms of the Agreement become the controlling legislation for the member states, and although they do retain a national patent office presence, they are to all intents and purposes sub-offices of the OAPI headquarters in Yaoundé. A single patent is granted by OAPI, carrying the "OA" code prefix, which has effect in all the member states, and (most significantly) the Organisation uses a centralised court structure which enables a single case to establish validity or order revocation for all the territories. The working language of the Organisation is French.

ARIPO

The English-speaking parallel organisation to OAPI is ARIPO, the African Regional Industrial Property Organization. Membership is determined by accession to the Lusaka Agreement of December 9, 1976. At the time of writing there are fifteen member states: Botswana, the Gambia, Ghana, Kenya, Lesotho, Malawi, Mozambique, Sierra Leone, Somalia, Sudan, Swaziland, Tanzania, Uganda, Zambia and Zimbabwe. Within the framework of OAPI, there are separate treaties on patents and industrial designs (the Harare Protocol of December 10, 1982, which entered into force on April 25, 1984) and on trademarks (the Banjul Protocol). All ARIPO states apart from Somalia have ratified the Harare Protocol.

Under this system, the applicant can file an application with either their national office or directly with the ARIPO office in Harare. As with other regional offices, a single application can have effect in all designated member states, but the ARIPO system–unlike OAPI and the EAPO–allows the applicant to designate fewer than the maximum number. The applicant uses only one language–English–and pays fees in U.S. dollars. There is also an advantage for the applicant in that they will only need to employ one agent. The system offers centralised processing of patent applications to grant and also administers renewals on behalf of the member states. Applications which are filed at a national office are forwarded to Harare for examination after the preliminary formal stages. After substantive examination is completed, copies of the application are sent to each designated state, which reserves the right to make a pre-grant declaration that the ARIPO patent will not have effect upon its territory.

In recent years, a revision of the Harare Protocol has made a link between it and the Patent Co-operation Treaty (see below). Under this system, designation of "ARIPO" in a PCT application is taken to mean automatic designation of all states party to both the Harare Protocol and the PCT. The country code for ARIPO is "AP" and the working language is English.

Nearly-Operational Systems

The Gulf Co-Operation Council (GCC)

The Patent Office of the Co-operation Council for the Arab States of the Gulf, to give it its full official title, is based in Riyadh, Saudi Arabia, and was founded in 1995. The implementing Regulations were passed in the following year and the operational Rules in 1998. It has six member states: Bahrain, Kuwait, Oman, Qatar, Saudi Arabia and the United Arab Emirates. The first patent applications were received in October 1998, but as yet, there does not appear to have been any publications issuing from the Office. As with the EPO, the GCC Office is not intended to replace national patent offices, although in the case of these countries, they did not all have a history of intellectual property prior to their membership of the GCC system–Oman and Qatar had no patent laws at all. The original term of the GCC patent was fifteen years, but this was revised in 2000 in order to bring the regional law into compliance with the TRIPS agreement, and now allows for a 20-year term from filing date.

At the time of writing, it is unclear whether any patents have in fact been granted by the GCC, and no country code appears to have been allocated for its publications. The table of contents of one issue of the GCC official gazette appears to suggest that applications are progressing through the office, but they have not appeared in any Western databases to date.

Ancom

The Andean Community is a group of South American states (Bolivia, Colombia, Ecuador, Peru and Venezuela) who have formed a common customs union for trade. The Community is guided by the Cartagena Agreement, which includes certain common provisions on industrial property. Their administrative organisation has issued a number of decisions flowing from this Agreement, the most recent in the patents area being Decision 486 on a "Common Intellectual Property Regime" dated September 14, 2000. This Decision establishes a common patent-granting framework for the member states. The new system will allow for 18-month early publication of patent applications, leading to a grant with a term of twenty years from filing. No publications have yet emerged from the system.

Systems Being Planned

In addition to the systems outlined above, there are a number of regional initiatives around the world which could lead to further co-operation in patent examination and granting. Amongst these are:

- The Arab League states (twenty-two members in North Africa and the Middle East), which would have a likely headquarters in Cairo, Egypt–the most recent amendment to Egyptian patent law allows for it to become the base for an Arab Patent Office,
- The Association of South-East Asian Nations (ASEAN) (ten members, founded in 1967), which has been working towards common intellectual property regulations, assisted by the EPO,
- Mercosur, the second 'common market' in South America (Argentina, Brazil, Paraguay and Uruguay, with Chile as associate member), which is investigating harmonization in the copyright area, but so far has made little progress in patents, and
- Caricom (the Caribbean Community), a grouping of fourteen countries in the Caribbean and parts of Central and South America.

Other trading blocs in Central America, South Asia, East and South Africa all have potential to form new systems of patent granting.

Other Systems

The Patent Co-Operation Treaty

The PCT is probably the most successful "regional patent system" of them all, if sheer numbers of applications filed is any measure. However, it is not a true regional system in that no patents are granted by the PCT. Instead, the operations of the PCT International Bureau, based in Geneva, allow for the streamlining of the early administrative phases of patent application, up to the publication at eighteen months. Thereafter, the applications are forwarded to designated national patent offices for further processing under national law. Unlike regional patent systems which grant a single document, a PCT international application matures into a cluster of national patents, indistinguishable from those which would have been obtained by individual applications in each country. The great benefit of the PCT is that it allows the applicant to defer the considerable cost of translation of their patent application into multiple official languages, until they have at least had a chance to assess the search report produced at 18 months (the so-called Chapter I procedure), often followed by an optional preliminary statement of patentability (Chapter II procedure). Documents published at eighteen months can appear in English, French, German, Japanese, Russian, Spanish or Chinese, and carry a "WO" code prefix to identify them.

Havana Agreement

This agreement was founded by the Council for Mutual Economic Assistance (CMEA), better known as the Comecon countries, during the Soviet period. Its members were the Soviet Union, Bulgaria, Cuba, Czechoslovakia, East Germany, Hungary, Mongolia, Poland, Romania and Vietnam. The organisation was founded in 1949 and dissolved in 1991. The purpose of the Havana Agreement was for mutual recognition of patent rights, but since few of the Communist states granted true patents, their publications did not appear very frequently. The most common manifestation of the system was the occasional East German patent with a front page in German but the body of the text in Russian. These inventions generally originated from the Soviet Union or

Czechoslovakia, and cited the appropriate priority details as part of the bibliographic record.

CONCLUDING REMARKS

This short survey of regional patent systems has described a very fluid area of legal activity. Although the World Trade Organisation has an interest in intellectual property, through the TRIPS agreement, there are still many bilateral and small multi-lateral trade agreements in the world today, with more being concluded each year. Whenever countries seek to strengthen trading ties, it brings with it an impetus to establish a genuine non-tariff trading system, with harmonised rights allowing for the free movement of goods and services between them. In the modern era, this desire for free trade is inseparable from a regime of effective intellectual property rights, and it seems inevitable that as more agreements come into operation, this will require a greater degree of uniformity in patent operations between member states. It seems unlikely that the days of the regional patent system are numbered, especially in the face of the outstanding success of the EPO and the very positive commencement of operations at the EAPO. For the searcher, it seems that information skills will have to co-exist with a reasonable legal knowledge and access to a good atlas, for the foreseeable future.

Patent Data for Technology Assessment, Part I: Applications, Patent Databases, and Retrieval Methods

Cynthia A. Kehoe
Xiao Jason Yu

SUMMARY. Patents have a wide range of uses for the clients of science, technology and business librarians, beyond legal patentability. This article discusses uses of U.S. patent information and compares several patent information systems. Specifically, the article highlights the value of patent information for technology intelligence and management; examines features of five major U.S. patent database systems; discusses critical retrieval issues; and provides examples of the types of queries for which patent databases are helpful. The article is the first of a two-part series; the second uses patent data to examine trends in global positioning system (GPS) technology, in order to more fully illustrate methods for technology assessment. *[Article copies available for a fee from The Haworth Document Delivery Service: 1-800-HAWORTH. E-mail address: <docdelivery@haworthpress.com> Website: <http://www.HaworthPress.com> © 2001 by The Haworth Press, Inc. All rights reserved.]*

Cynthia A. Kehoe, PhD, MLIS, is Independent Consultant, P.O. Box 3442, Urbana, IL 61803-3442 (E-mail: cynthiakehoe@yahoo.com). Xiao Jason Yu, MA (Social Anthropology), MS in LIS, is Information Core Director, California Center for Population Research, University of California, Los Angeles, CA (E-mail: xjyu@ccpr.ucla.edu). This article was written during his tenure as Data and Electronic Services Librarian, Indiana University Libraries, Bloomington, IN.

[Haworth co-indexing entry note]: "Patent Data for Technology Assessment, Part I: Applications, Patent Databases, and Retrieval Methods." Kehoe, Cynthia A., and Xiao Jason Yu. Co-published simultaneously in *Science & Technology Libraries* (The Haworth Information Press, an imprint of The Haworth Press, Inc.) Vol. 22, No. 1/2, 2001, pp. 101-116; and: *Patent and Trademark Information: Uses and Perspectives* (ed: Virginia Baldwin) The Haworth Information Press, an imprint of The Haworth Press, Inc., 2001, pp. 101-116. Single or multiple copies of this article are available for a fee from The Haworth Document Delivery Service [1-800-HAWORTH, 9:00 a.m. - 5:00 p.m. (EST). E-mail address: docdelivery@haworthpress.com].

http://www.haworthpress.com/store/product.asp?sku=J122
© 2001 by The Haworth Press, Inc. All rights reserved.

KEYWORDS. Patents, patent databases, technology assessment, research methodology

INTRODUCTION

This article discusses applications of U.S. patent information, compares several patent information systems, and identifies some critical retrieval issues. It is the first of a two-part series; the second article uses the patents issued for the technology of global positioning systems (GPS) as an example to illustrate how patent data can be retrieved and analyzed to find patterns of development and innovations in an area of technology (Yu and Kehoe 2001).

Patent databases are perhaps most often thought of as sources for legal research–answering questions such as "can this device I've invented be patented?" However, patents have a wider range of uses for business and science-technology librarians and their clients. For example, patents can be used to track the direction in which a technology is developing, or to identify the leading players in a field. Other applications are in strategic planning, R&D and product development, and other business and economic applications. Though other articles have argued for the value of patent data for business intelligence (Mogee 1997, Vijay-Rao 2001), few have provided detail on selecting systems and methodology for retrieval and analysis. This series attempts to address that gap.

Many organizations are now concerned with intellectual asset management. Technology transfer is mandated for universities and government agencies. Development firms do initial research, then sell or license technologies elsewhere. Firms that actively patent may develop core technologies, and license others. For these organizations and their customers and competitors, patent information retrieval and analysis is critical.

Patents can be useful for strategic planning and technical intelligence. Among the many applications for patent documents, a business may use them to:

- Monitor worldwide technological developments;
- Determine emerging trends in an industry;
- Obtain state-of-the-art technology information;
- Locate expert consultants or potential employees;
- Find potential joint venture partners or merger/acquisition opportunities;

- Identify product licensing opportunities;
- Identify competitors and watch for emerging competitors;
- Monitor the activities and plans of competitors, their R&D focus; and
- Analyze investment opportunities.

Those in manufacturing and R&D may use patents to determine whether to pursue an idea, avoid duplication in research, or look for manufacturing problem solutions.

Patents are resources for historians of industry and technology, economics and economic development specialists. Patents have been used to:

- Study innovation trends, over time or by nation;
- Examine the impact of a company, or the growth of a technology; and
- Monitor the international diffusion of a technology.

While using patents to examine technologies and industries does require some knowledge of the content of patent documents, it is not as complex as legal patentability searching. Studies may focus on the patents of a particular organization, inventor, or geographic region. More complex analyses focus on a particular type of technology, using the classification scheme which provides subject access to U.S. patents. This necessitates learning to identify the appropriate classes and subclasses, with the help of classification tools developed by the U.S. Patent and Trademark Office (USPTO): the *Manual of Patent Classification*, its *Index* and *Definitions*. These finding aids are available in searchable versions on the USPTO Web site, and a number of books on patent searching (Ardis 1991; Wherry 1995) have sections on using the U.S. patent classification system. For basic information about U.S. patents, see the Nolo Press *Legal Encyclopedia* online (Nolo Press), or the USPTO publication, *General Information Concerning Patents* (USPTO).

VALUE OF PATENTS

Patents are a public record of research and development activities, as conducted in corporations, universities, government agencies, and by individual inventors. Over 35 million patents have been granted worldwide. Approximately 80% of the information in patents is not published elsewhere (PATSCAN 2002). Patenting is on a country-by-country basis. Because the United States is a major technology market, technologies that their inventors believe are important will be filed in the United States, though the inventors and organizational patentholders may reside elsewhere. Patents granted to firms outside the United States represent a large portion of those in the U.S. patent system. According to patent statistics released by the USPTO (2002), the top ten patenting organizations for 2001 consist of two U.S. corporations, seven Japanese corporations, and one corporation from the Republic of Korea. Patents granted in the United States represent key technological inventions from around the world. Under the U.S. patent system, these technologies are described in English, and patents are required to contain enough detail to enable an expert to recreate the technology.

Because of the patent laws regarding disclosure, patent applications are typically the first public disclosure of a technology. Patents therefore can provide an early alert concerning new directions in the development of a technology, new directions in R&D focus for a particular firm, or potential new competitors. Patent applications have long been public documents in Europe, but in the United States this is a recent event. Beginning in March 2001, applications that are 18 months old (with some exceptions) are now available, in addition to the patents granted. Patents are indicators of technological output, and indicators of technology direction. Highly cited patents are technologically important and economically important, and firms with highly cited patents have greater market value (Trajtenberg 1990; Mogee 1997; Hall et al. 2000).

A U.S. patent typically contains citations to five or six U.S. patents in a section labeled References Cited. This "bibliography" is created by the patent examiner rather than the inventor, and may also include sections of foreign patents and literature. These patent-to-patent citations identify the prior art on which the new patent builds. Studies have used them to show knowledge links between firms. Highly cited patents tend to be those that contain high-impact discoveries, not just minor variations on old inventions (Hall et al. 2000; Jaffe et al. 2000). This more advanced technique is not a focus of this paper.

Patents as Technology Development Indicators

The U.S. Patent and Trademark Office grants U.S. patents and provides access to the resulting U.S. patent documents. Access to the information contained in patents is, in fact, the reason for the existence of a patent system. Inventors are given a limited monopoly in exchange for sharing new knowledge, so as to accelerate the pace of technology development and thus the economic development of the nation. That about 80% of the information contained in patents is not published elsewhere by the inventors is one indication of the need for such an incentive. Access is through the products provided by the USPTO itself, both print and electronic, and through commercial information providers to whom the USPTO has sold the data. The USPTO has always attempted to provide access to patent data for people who are not expert researchers, but the development of its own electronic tools has lagged behind the products that commercial information services have been able to provide (Wherry 1999). The Web has narrowed that gap, at least for simple queries.

Patents have long been seen as technology development indicators. As pointed out by Hall et al. (2000), the scholarly use of patent data in the analysis of technological change stretches back to the path-breaking analyses of Schmookler (1966) and Scherer (1965). According to Trajtenberg (1990), patents have exerted a compelling attraction on economists dealing with technological change, particularly since Jacobs Schmookler's seminal work in this area (1966).

The availability of patent information from the USPTO in computer-readable form in the late 1970s prompted greater interest in econometric analyses using this patent data; much of this work is reported in Griliches' review article (1984). In the late 1980s, patent citation information began to be available in computerized form, which led to another wave of economics-based patent research, and allowed investigation of an additional set of questions related to the flow of knowledge across time, space and organizational boundaries. Patent data provide rich technological, geographic and institutional detail, and are publicly available for all kinds of research institutions (firms, universities, other non-profits, and government labs) in virtually every country (Hall et al. 2000; Trajtenberg 1990).

Patent analysis is not only U.S.-based. Patent data is highly valued in Europe and other areas as well. In 1994 the Organisation for Economic Cooperation and Development (OECD), a European-based inter-governmental organization, published a working manual titled *Using Patent Data as Science and Technology Indicators* (1994). The manual noted that "In recent years analysts and policymakers have made increasing use of patent indicators to analyse the rate and direction of technical change. The requirement of novelty for the granting of patents means that the indicators are particularly appropriate for advanced countries; they may not adequately portray technological activity in less developed countries." OECD also recognized that the reliability of patent data as an indicator of technological innovation is illustrated by a number of surveys, showing that a large proportion of firms' inventions are patented and that a large proportion of patents become innovations with an economic use. Moreover, as pointed out by this OECD document, the patents give a good picture of invention and innovation in both small and large firms, something that other common R&D indicators do not properly measure, since they are largely drawn from data from public companies (OECD 1994).

Patent Data for Technology Intelligence

Patents can be a resource to acquire information on the innovative activities of firms (Archibugi et al. 1996, Ardis 1991, Albert and Hicks 1998), or a technological forecasting tool for relevant firms (Campbell and Levine 1984). Mogee (1997) and Vijay-Rao (2001) have explicitly treated patent analysis as a means of corporate competitive intelligence. CHI Research is a business intelligence firm that analyses patent data, and its Web site is a good resource for exploring the variety of types of studies that may be conducted. Mogee (1997) states that "Patents are an important source of technological intelligence that companies can use to gain strategic advantage." She notes that companies use patent information for competitive intelligence in two key ways. The more common way is as a current awareness tool. Many firms circulate patent abstracts to researchers and other technical people to ensure that they keep up to date on developments in their field. The second way in which patents are used involves statistical analysis of large numbers of patents to discover broad patterns or trends that may be significant to the firm's technology management or strategy. Such patent analyses produce quantitative results that can be used in conjunction with technology intelligence methods based on expert opinion, which are prone to problems of bias and information inadequacy. Similarly, Vijay-Rao (2001) treats patents as valuable sources of information for corporate competitive intelligence with a focus on firms' technological strengths.

PATENT DATABASES AND RETRIEVAL ISSUES: A SCENARIO-BASED COMPARATIVE REVIEW

Most countries have a patent system, and many have created databases for public use, and/or include their records in international databases such as *Derwent World Patents Index*. With the advent of the Web, patent information is much more readily available. In addition to national patent databases, there are subject-focused ones on the Web such as the *DNA Patent Database*. As well, in certain fields in which major research is published in the form of patents, the literature databases index patents–e.g., *Chemical Abstracts* (CAS).

The sci-tech librarian who is not an expert patent searcher is not likely to use one of the specialized patent systems such as that of Questel-Orbit, which requires a certain level of specialized search knowledge. Following is a comparison of some of the systems that a librarian who is an occasional patent searcher is more likely to choose from among: the USPTO Web database; the USPTO DVD collection; Delphion on the Web; Dialog's *U.S. Patents Fulltext*; or *USPAT* from LexisNexis.

The types of patent analyses conducted might focus on inventors, assignees, fields of technology, or technology developments by region or over time. Assignees are the organizations that hold the patents–most often the employers whose personnel carried out the research. Corporations, government agencies, universities and other types of organizations may hold patents. While all patents have inventors, not all patents have assignees. The criteria used to compare the five systems focus on the capabilities needed to do such analyses. As well, sample scenarios are provided to help demonstrate some of the typical problems and techniques.

The initial task is to understand the problem and match it to the appropriate database. The primary criteria to choose a source are the fields to be used in the query, and the fields needed for analysis. One might, for example, search by subject class, but analyze by assignee and date. Additional factors include the ease with which the needed fields can be downloaded, analysis tools such as sorting, and the manner in which the information provider treats those fields with content than can change, which may necessitate revision of older patent records.

One of the challenges of maintaining a patent database is that certain information in an individual patent can change. The two key fields for the types of analysis under discussion here are assignee and classification. A patent may be granted to an individual inventor, who later sells it to a firm, for example, or one company may acquire another, including its intellectual property, and patent ownership changes are recorded at the USPTO. Some information service providers have chosen not to update the assignee field, but to keep the assignee at the time the patent was granted. For most types of analysis, this is useful, since it reflects the organization that carried out the R&D; but there are times when current ownership is important.

The classification system also undergoes regular revision to reflect new technologies. When changes are made, older patent records are changed in the USPTO system, so that a search by the current class will retrieve all relevant patents. For classification changes, the providers do attempt to update older patent records. The accuracy of changes and special quirks of searching are noted below.

The availability of types of search queries and formatting and downloading features are also shown in Tables 1 and 2, to allow quick comparison.

USPTO Web

The USPTO's Web-based patent database allows one to search by inventor, assignee (organizational patentholder) at the time the patent was granted, keyword in any field of the patent, current classification (field of technology), geographic location of inventor or assignee, and by date patents were filed or granted (among other fields). It also supports simple citation searching–tracking what earlier patents or publications a patent cites, or what later patents cite the one of interest. The search capabilities are excellent, though the interface is not always easy to use.

This is a good database for an inventor to explore when looking for solutions to a manufacturing problem or examples of the drawings used in patents, especially when time is not a serious constraint. But its limited format and downloading options do not make it the best choice for most analyses. One can either download a list, in order of date granted, that lists patent number and title, or one can download individual patents in full text. Downloading a set of patents is not possible, nor is selecting the fields desired, or sorting the records. These constraints are a deliberate part of the design, so as to discourage firms that do extensive patent searching from overloading the USPTO site.

The USPTO site is useful for certain types of trend analysis. Example: How many patents were granted to organizations headquartered in Austin, Texas, over a ten-year period? Has innovative activity increased in the last decade? Perhaps an economic development agency wants to use such information to attract other businesses. This type of query primarily needs counts rather than specific patent records. The USPTO Web database is also useful for testing initial queries before implementing them in commercial systems. As well, the site provides the *Manual of Classification*, and its *Index* and *Definitions*, in a searchable format, to allow the user to identify the appropriate classification terms. This is an excellent source for the classification finding tools, even when the search is to be conducted elsewhere.

USPTO Patents BIB DVD

The USPTO issues a bibliographic database of U.S. patents in a DVD format–formerly the *Cassis BIB* CD-ROM. This database is provided to all patent depository libraries; other organizations may subscribe at a relatively low cost. The product is updated every two months–unlike the other databases mentioned here, which are updated weekly. Whether currency matters therefore affects your choice. This is part of a family of databases, but the focus here is on the bibliographic database alone. (For information about the other products, see the product catalog issued by the USPTO, available at its Web site.) Search capabilities are more limited in the *Patents BIB* DVD than for the Web database. One cannot search by (or view) inventor name. Keyword searching is limited–only the most recent years of records include an abstract, and title searching in patent databases is not very useful because the titles are often not very descriptive. Subject searching in the *Patents BIB* database must be done with the use of the classification system, which requires more search knowledge. One can also search by assignee at the time the patent was granted, geographic information and date granted. Citation searching is not available. The database does support sorting, downloading a group of records, and selecting the fields in the record to download.

Trend analyses such as described above for innovative activity by Austin-based firms can be easily done with the *Patents BIB* database, and can be extended to incorporate information about which organizations were active in what periods. One can easily download the needed information and further manipulate it in a database or spreadsheet if needed. Other query examples: Are most roses developed by individual growers or do the plant patents have assignees when they're granted? Searching by class and then sorting by assignee will answer this query. Are flying discs, of which Frisbees are one brand, still being granted patents? That is, is this still an active field of product development? Is most of this activity by individuals or companies? How international is it? These queries require use of the patent classification.

Delphion

The Delphion Web site has free and fee-based sections. Delphion provides free access to the U.S. patent database, with some search and display limitations, and it is this resource on which we will focus. A query searches the text of the 'front page' of a patent–in effect, the title, abstract, inventor and assignee (at the time the patent was granted). One can limit by pre-established ranges of years. One cannot search by class or do citation searching. (The fee-based version supports much more sophisticated search and data analysis capabilities.) Like its USPTO Web counterpart, downloading of groups of records (except as a citation list) is not supported. However, Delphion does support some customization of the formatting for patent lists, which can be very useful for queries based on inventor or original assignee. It allows a subject search which can be expressed by keyword (and for which you need not be as concerned about comprehensiveness as one is in legal patentability queries).

A simple scenario: A university technology management office needs to identify potential licensees for an invention of one of the university faculty. The U.S. patent database identifies firms investing in R&D in this field. More specifically, the invention is a drug treatment for sleep apnea. A keyword search of the titles and abstracts, in the U.S. patent database provided by Delphion, retrieves about 175 patents for the most recent five years that mention sleep apnea, and less than 30 concerning drug or pharmacological approaches. The default display format is patent number, date granted, and title, presented in date or in relevancy order. However, this can be customized. Choosing from a set of options at the bottom of a results page specify that one wishes to see the patent number, date granted, assignee, and title. Then sort by assignee. The result is an easily printed list showing the number of patents held by each organization. A quickly executed strategy using a free resource provides some useful information–in this case, one finds that Eli Lilly and the University of Wisconsin have been conducting research in this area. There are fee-based patent services that support more complex strategies and provide more analysis and customization tools–but they are not needed for this straightforward query.

USPAT from LexisNexis

The information provider LexisNexis has numerous online products, including the *LexisNexis* system, *lexis.com* and *LexisNexis Academic Universe*. Many of its online products, including the three named, have the U.S. patent database (*USPAT*) in full text. Complex search strategies and customized display formats are available, though in more limited ways for *Academic Universe*. However, potential problems with data quality may limit the usefulness of this resource. Assignee and classification are two key fields which are updated–patents change ownership or class numbers are revised. In *USPAT* these fields have not been accurately updated in the past (Kehoe 2000). The U.S. patent files were reloaded in June 2002, to incorporate enhancements. The data was therefore correct again as of that date. But the database must be monitored to determine whether the previous problems–whether caused by flaws in the procedures or in the update software–have also been corrected. Assuming that past problems have been corrected, *USPAT* and *U.S. Patents Fulltext* on Dialog are the only databases among these five systems that does track changes in the assignee field, making it quite valuable if one's question is about the current owner of a patent rather than which organization carried out the research.

If one has queries that do not rely on the current classification or the most recent assignee ("assignee after issue"), either for the search or analysis, then the ability to download a set of records, and select the fields to be downloaded, makes *LexisNexis* or *lexis.com* a useful option. Combining several criteria in one's search is easier with a more complex system like *lexis.com*–once one learns the field labels and format for entering queries. Searchers of *LexisNexis Academic Universe* can use most of the field labels available in the more full-featured systems, though this is not well documented. *Academic Universe* also supports downloading a set of records in a brief format, but does not allow the searcher to select the fields to be downloaded. Downloading is not in a format easily imported into a spreadsheet.

Scenario examples: The doctoral programs of the University of Illinois and the University of Texas are both highly rated in chemical engineering. Does this extend to the number of patents in this field that each has received?

U.S. Patents Fulltext on Dialog

This database requires the most complex search knowledge. However, it also allows the most complex formatting of results, and provides analysis tools such as sorting and ranking. The patent database supports searches for inventor, assignee at issue or after issue, classification, geography, date, citation searching, etc. Unlike the LexisNexis *USPAT* database in the past, the U.S. class field accurately reflects the current classification. The trade-off is that this may be the most expensive choice, depending on one's subscription plan. Therefore it behooves the librarian to learn when to choose the USPTO or Delphion products.

Scenario example: A client needs to know what patents are currently held by a firm. The company has been through a period of merger and acquisition activity. Therefore, knowing the organizational assignee when the patent was granted is not adequate. Dialog allows the searcher to identify current assignees. One can then rank the records based on the original assignee, to identify the sources from which the firm of interest acquired its technologies. (Note: Although organizations are supposed to report changes in ownership of patents to the USPTO, this does not always occur.)

Comparisons

Tables 1 and 2 allow the reader to compare the features of the five systems discussed herein. Table 1 examines the availability of different types of queries. Table 2 examines the ability to download batches of records, in a format useful for analysis.

Historical Analysis

The USPTO Web database supports searching of numerous fields for patents issued since 1976. Patents issued from 1790 through 1975 are searchable only by patent number and current U.S. classification. These older patents can be downloaded or printed only as images, or as lists of patent numbers with their assigned classes. The USPTO *Patents BIB* DVD contains bibliographic information for utility patents issued from 1969 to the present, and for plant and design patents issued from 1977 to the present. The Delphion *U.S. Patents Granted* database, *USPAT* from LexisNexis, and *U.S. Patents Fulltext* from Dialog all begin with patents issued in 1971.

TABLE 1. Availability of Query Type

	USPTO Web	USPTO Patents BIB	Delphion (free section)	LexisNexis, including Academic Universe	Dialog
Inventor	Yes	No	Yes	Yes	Yes
Assignee at Issue	Yes	Yes	Yes	Yes	Yes
Geography	Yes	Limited	Limited	Yes	Yes
Date Granted	Yes	Yes	Limited	Yes	Yes
Date Filed	Yes	No	No	Yes	Yes
U.S. Class	Yes	Yes	No	Problems	Yes
Text Search	Full patent	Limited	Front page	Full patent	Full patent
Citation Searching	Yes	No	No	Yes	Yes

TABLE 2. Availability of Formatting and Downloading Options

	Sort results	Select fields to display or download	Download set of records (other than citations)
USPTO Web	No	No	No
USPTO Patents	Yes	Yes	Yes
Delphion (free)	Limited	Limited	No
LexisNexis	No	Yes	Yes
Academic	No	No	Yes
Dialog	Yes	Yes	Yes

CONCLUSION

Advances in information technology have greatly improved access to patent data. As a result, the uses of patent information have expanded enormously. In the early years, uses for patents were primarily concentrated on protection, R&D, and product design solutions. However, with the greater ease of retrieving and using patent data, patents have become more important as a source of information for strategic planning and other business applications. Better patent retrieval systems, with the inclusion of citation data, have stimulated new scholarly ventures in which patent records are extracted and mined to explore the history of a technology, or innovation trends or patterns of economic development, over time and across geographic boundaries.

For sci-tech librarians, the evolving patent systems and applications bring about both opportunities and challenges. On the one hand, the localized and distributed access to patent data through compact storage technology and computer networks invites more clients to sci-tech librarians, supports more complex queries, and may make their work more efficient. On the other hand, the variety of uses and the availability of multiple patent database systems require that sci-tech librarians have good knowledge of the pros and cons of major database systems and appropriate retrieval methods associated with them, and the ability to answer end user questions concerning complex systems. To help sci-tech librarians to meet such challenges, we have presented a comparison of the features of five major databases of U.S. patents, namely, the USPTO Web patent database, the USPTO *Patents BIB* DVD, Delphion's *U.S. Patents Granted*, the LexisNexis U.S. Patents file (*USPAT*), and Dialog's *U.S. Patents Fulltext*. We have also identified retrieval issues associated with particular databases and applications. As the examples illustrate, each of the five patent database systems has advantages for particular types of queries, and searchers are guided in selecting systems with appropriate database features for specific applications.

REFERENCES

Albert, Michael B. and D. Hicks. 1998. Using Patent Data for Strategic Planning. *SATM Innovation & Technology Management News*, 2 (2), Fall/Winter 1998.

Archibugi, D. et al. (1996). Innovation Surveys and Patents as Technology Indicators: the State of the Art. In: Innovation, Patents, and Technological Strategies, 17-56. Edited by OECD. Paris: OECD Press.

Ardis, Susan A. 1991. *An Introduction to U. S. Patent Searching: The Process.* Englewood, CO: Libraries Unlimited.

Campbell, R. S. and L. O. Levine. 1984. *Technology Indicators Based on Patent Data: Three Case Studies.* Richland, WA: Battelle Pacific Northwest Laboratories.

CAS. *Chemical Abstracts.* American Chemical Society. Available at: http://www.cas.org/prod.html. Accessed November 2002.

CHI Research, Inc. Company Web site available at http://www.chiresearch.com. Accessed November 2002.

Delphion, Inc. *U.S. Granted Patents.* Available at: http://www.delphion.com/. Accessed November 2002.

Derwent World Patents Index. Thomson Derwent. Available at: http://www.derwent.com/worldpatentsindex/index.html. Accessed November 2002.

DNA Patent Database. Georgetown University, Kennedy Institute of Ethics and the Foundation for Genetic Medicine. Available at: http://www.geneticmedicine.org/fp_dpd.htm. Accessed November 2002.

Griliches, Zvi. 1984. *R&D, Patents and Productivity.* Chicago: University of Chicago Press.

Hall, Bronwyn H., Jaffe, Adam B. and Manuel Trajtenberg. 2000. *Market Value and Patent Citations: A First Look.* Cambridge, MA: National Bureau of Economic Research, NBER Working Paper, No. 7741.

Jaffe, Adam B., Manuel Trajtenberg and M. S. Forgarty. 2000. Knowledge Spillovers and Patent Citations. *American Economic Review* 90(2): 215-218.

Kehoe, Cynthia A. 2001a. *Major Players in the International Rose Market: Using Patent Analysis to Examine an Industry.* Working Paper ISRN UIUCLIS-2001/3+INFO, Graduate School of Library and Information Science, University of Illinois at Urbana-Champaign.

_____. 2001b. *Patents for Business Intelligence.* Presentation, National Technological University, February 2001.

_____. 2000. Quality Patent Searching. Presentation, *National Online Meeting 2000.* New York: Information Today, Inc.

Mogee, Mary Ellen. 1997. Patents and Technology Intelligence. In: *Technical Intelligence for Business: Keeping Abreast of Science and Technology*. Edited by W. B. Ashton and R. A. Klavans. Washington, DC: Battelle Press.

Nolo Press. Patents. In: *Legal Encyclopedia*. Nolo Press Law Center Available at: http://www.nolopress.com/lawcenter/. Accessed November 2002.

Organization for Economic Cooperation and Development. 1994. *Using Patent Data as Science and Technology Indicators: Patent Manual*. OECD Working Papers, 2 (66). Paris: OECD.

PATSCAN, Walter C. Koerner Library, University of British Columbia. *Why Search Patents?* Available at: http://www.library.ubc.ca/patscan/whysearc.html. Accessed November 2002.

Questel-Orbit. Available at: http://www.questel-orbit.com/. Accessed November 2002.

Scherer, M. 1965. Firm Size, Market Structure, Opportunity, and the Output of Patented Inventions. *American Economic Review*, 55(4): 1097-1125.

Schmookler, J. 1966. *Invention and Economic Growth*. Cambridge, MA: Harvard University Press.

Trajtenberg, M. 1990. *Economic Analysis of Product Innovation: The Case of CT Scanners*. Cambridge, MA: Harvard University Press.

U.S. Patent and Trademark Office (USPTO). *Classification Definitions*. Washington, DC: U.S. Government Printing Office. Also available at: http://www.uspto.gov/web/patents/classification/. Accessed November 2002.

_____. *General Information Concerning Patents*. Available at: http://www.uspto.gov/web/offices/pac/doc/general/index.html. Accessed November 2002.

_____. *Index to the United States Patent Classification System, December 2001*. Washington, DC: U.S. Government Printing Office, 2002. Also available at: http://www.uspto.gov/go/classification/uspcindex/indextouspc.htm. Accessed November 2002.

_____. *Manual of Classification*. Washington, DC: U.S. Government Printing Office. Also available at: http://www.uspto.gov/go/classification/. Accessed November 2002.

_____. *Patent Full-Text and Full-Page Image Databases*. Available at: http://www.uspto.gov/patft/. Accessed November 2002.

_____. 2002. *Preliminary List of Top Patenting Organizations: Calendar Year 2001*. Available at: http://www.uspto.gov/web/offices/ac/ido/oeip/taf/top01cos.htm. Accessed November 2002.

_____. *Products and Services Catalog 2001*. Available at: http://www.uspto.gov/web/offices/ac/ido/oeip/catalog/index.html. Accessed November 2002.

U.S. Patents Fulltext. Dialog Corp. Available at: http://library.dialog.com/bluesheets/html/bl0654.html. Accessed November 2002.

USPAT, *U.S. Utility, Design and Utility Patents*. LexisNexis. Available at: http://web.nexis.com/sources/scripts/info.pl?228804. Accessed November 2002.

Vijay-Rao, Geeth. 2001. Creating Intelligence from Data: Resources and Techniques for Leveraging Patent Information for Competitive Advantage. In *National Online Meeting Proceedings*–2001, 505-534. Edited by Martha E. Williams. New York: Information Today, Inc.

Wherry, Tim L. 1999. Patents in the New World. *Science & Technology Libraries* 17(3/4): 217-22.

_____. 1995. *Patent Searching for Librarians and Inventors*. Chicago: American Library Association.

Yu, Xiao J. and Cynthia A. Kehoe. 2001. Patent Data for Technology Assessment, Part II: Using U.S. Patent Data to Examine the Trends of GPS Technology. *Science & Technology Libraries* 22(1/2): 117-135.

Patent Data for Technology Assessment, Part II: Using U.S. Patent Data to Examine Trends in GPS Technology

Xiao Jason Yu
Cynthia A. Kehoe

SUMMARY. This case study of the use of U.S. patents for technology intelligence is the second article of a two-part series on applications of U.S. patent data. Sci-tech librarians are often drawn into the intersection with business information research that involves the study of technology trends and technical industries for business decision-making. Patent data can be a useful resource for solutions to such information problems. This paper uses the patents issued in the technology of global positioning systems (GPS) as an example to discuss how patent data can be retrieved

---

Xiao Jason Yu, MA (Social Anthropology), MS in LIS, is Information Core Director, California Center for Population Research, University of California, Los Angeles, CA (E-mail: xjyu@ccpr.ucla.edu). This article was written during his tenure as Data and Electronic Services Librarian, Indiana University Libraries, Bloomington, IN. Cynthia A. Kehoe, PhD, MLIS, is Independent Consultant, P.O. Box 3442, Urbana, IL 61803-3442 (E-mail: cynthiakehoe@yahoo.com).

The authors would like to thank Mary F. Popp, Public Services Librarian of Library Information Technology, Indiana University Libraries, for her invaluable input and suggestions on an earlier draft of this paper.

[Haworth co-indexing entry note]: "Patent Data for Technology Assessment, Part II: Using U.S. Patent Data to Examine Trends in GPS Technology." Yu, Xiao Jason, and Cynthia A. Kehoe. Co-published simultaneously in *Science & Technology Libraries* (The Haworth Information Press, an imprint of The Haworth Press, Inc.) Vol. 22, No. 1/2, 2001, pp. 117-135; and: *Patent and Trademark Information: Uses and Perspectives* (ed: Virginia Baldwin) The Haworth Information Press, an imprint of The Haworth Press, Inc., 2001, pp. 117-135. Single or multiple copies of this article are available for a fee from The Haworth Document Delivery Service [1-800-HAWORTH, 9:00 a.m. - 5:00 p.m. (EST). E-mail address: docdelivery@haworthpress.com].

http://www.haworthpress.com/store/product.asp?sku=J122
© 2001 by The Haworth Press, Inc. All rights reserved.

and mined to find patterns of development and innovations in an area of technology. GPS technology is found to be a rapidly emerging and changing field, dominated by a few American corporations and geographically concentrated, but of interest to many organizations in different fields and a number of countries. *[Article copies available for a fee from The Haworth Document Delivery Service: 1-800-HAWORTH. E-mail address: <docdelivery@haworthpress.com> Website: <http://www.HaworthPress.com> © 2001 by The Haworth Press, Inc. All rights reserved.]*

KEYWORDS. Patents, patent analysis methodology, patent data mining, technical intelligence, technology assessment, global positioning systems (GPS)

INTRODUCTION

Sci-tech librarians are often drawn into the intersection with business information research that involves the study of technology trends and technical industries for business decision-making. As well, organizations are managing their patents as intellectual assets, and licensing technologies that are not central to their business focus. Patent data can be a useful resource for trend analysis, technology directions and R&D trends, and the identification of licensees for technology transfers or merger/acquisition targets. This paper uses the patents issued in the technology of global positioning systems, or GPS, as an example to discuss how patent data can be retrieved and mined to find patterns of development and innovations in an area of technology.

Although a number of articles have argued the need to include patents as a source of information for literature searches and market research on technology topics (Albert 1998; Mogee 1997; Vijay-Rao 2001), it is much less common to discuss the techniques for doing so. Aiming to fill such a gap, methodology is a major focus of this article. In the first article of our two-part series, we argued the value of U.S. patent data for certain kinds of sci-tech and business information research, compared several database systems often used by people who are not expert patent searchers, and provided a set of search scenarios (Kehoe and Yu 2001). This second article focuses in more detail on the questions that one might pose against patent data, methodologies for addressing them, and types of findings, using GPS-related patents as a case study in technology assessment.

Combining Trajtenberg's (1990) technological field approach with the firm-as-unit analysis method (Griliches 1984), this study uses patents as data to examine the technological trends of a specific field—Global Positioning System (GPS)—and to conduct an analysis of the relevant firms' performance in this field.

GPS AS A FIELD OF TECHNOLOGY DEVELOPMENTS

The Global Positioning System (GPS) is a worldwide, all-weather, satellite-based navigation system. This network of satellites continuously transmits coded information, which makes it possible to precisely identify locations on earth by measuring distance from the satellites. The satellite system began as a U.S. Department of Defense effort with the first launch in 1978, and was completed in 1994 with the deployment of the 24th satellite, creating a virtual net of satellite coverage over the entire globe. GPS now has a variety of commercial applications, often grouped into five categories: location, navigation, tracking, mapping and timing. Recreational use has been growing rapidly (Garmin International; Trimble Navigation).

Since 1984, when the first commercial GPS receiver was put on the market, sales of GPS technology have increased sharply. By 1991, worldwide sales revenue of receivers was about $300 million; by 1995 it was approximately $700 million, and by 1997, almost $2 billion. The worldwide market for GPS products and services is predicted to reach at least $14 billion by the year 2005. Gains are being seen in all markets, including in-vehicle navigation, communications equipment, vehicle/freight tracking, agriculture, aviation, marine systems, and surveying and mapping equipment (Steede-Terry 2000). This field produces incessant technological innovations, as its markets and applications continue to develop. Because innovation is occurring rapidly and across a number of industries, patent documents can be a key source for learning about new developments. Technologies are disclosed in patent records before they get to market. Developments that are components of other devices may not be obvious to market observers, but may be granted individual patents. This study uses U.S. patent data to analyze certain trends of this field. The study demonstrates how data analysis can transform the raw data of patent records into valuable information and intelligence for corporate decision-making and technology development.

METHODOLOGICAL ISSUES

Why Choose U.S. Patent Data?

Most countries have a patent system, and many have created databases for public use, and include their records in international databases such as *Derwent World Patents Index*. The U.S. patent system reflects the world's biggest technology market, and captures indicators of international technology diffusion and transfer from other nations. Patenting is on a country-by-country basis, but assignees outside the U.S. tend to file key patents in the United States. GPS technology also has strong roots in the United States, beginning as an activity of its military.

There are also trade-offs in using a non-U.S. patent database. For example, some international patent databases would require identifying equivalent patents filed in different countries, in order not to give too much weight to duplicate patents. Using a single country database avoids this, when a full global perspective isn't essential, but is sometimes too narrow. However, based on the U.S. foundation of this technology, the study uses U.S. patent data as the source to illustrate the usefulness of patent analysis and to examine GPS technology developments.

Data Retrieval

In retrieving patent data for this study, we followed the strategy outlined by Kehoe (2001a). Using the classification tools available on the U.S. Patent and Trademark Office (USPTO) Web site, the patent classes and subclasses for GPS technologies were identified. The primary subject access to U.S. patents is by classification.

The USPTO grants U.S. patents, gives access to the resulting patent documents, and distributes patent information in electronic form to other database providers. The USPTO does not have a subject heading scheme for patents (though some information providers add their own subject terms). Rather, the USPTO has developed a classification scheme, which is documented in the *Manual of Classification*, its *Index*, and *Definitions*, available in searchable form on the USPTO Web site. This Web site provides good documentation and search aids, whether one is using a USPTO database or the U.S. patents as provided by other information vendors.

The initial task is to understand the problem and match it to the appropriate database. In part one of this series, five systems were compared–

the USPTO Web site, the USPTO *Patents BIB* DVD, the *U.S. Patents Granted* free section of the Delphion Web site, U.S. patents on LexisNexis (*USPAT*), and *U.S. Patents Fulltext* on Dialog. This study of a technology requires searching by class and date, and downloading results in an extended bibliographic format that includes the inventor and the organization to which the patent was issued (the assignee). Among these five choices, Dialog best fits these criteria.

For analysis, one might choose to transfer data files into a bibliographic database such as ProCite or EndNote; a spreadsheet such as Excel (used for this analysis), that supports analysis and allows results to be displayed with statistical conventions and graphics; SPSS data files for further statistical analysis; or DBF files as an attribute table for a geospatial analysis with geographical information systems (GIS) software. For small sets of results, reviewing and sorting the records in a word processor is often adequate.

SEARCH STRATEGIES AND ANALYSIS

Using the classification tools at the USPTO Web site, one finds that GPS patents are assigned to two classes: Class 342 is titled Communications: Directive Radio Wave Systems and Devices; and Class 701 is titled Data Processing: Vehicles, Navigation, and Relative Location. The several subclasses in each class provide one picture of the way the field can be divided from a technical viewpoint. The classification system is hierarchical. In class 342, subclasses 350, 352 and 357.01 are provided here to give context–the relevant GPS subclasses begin with 357.06. The dots preceding various headings indicate the hierarchy: Directive is a major category; one type of directive communication system includes a satellite; some satellite systems have position indicating capabilities (357.01), and a subset of this group is those that use GPS to do so (beginning with 357.06). In class 701, the four relevant subclasses are subsets of navigation devices (701/200) that employ position determining equipment (701/707).

Class 342 COMMUNICATIONS: DIRECTIVE RADIO WAVE SYSTEMS AND DEVICES

(E.G., RADAR, RADIO NAVIGATION)

350   DIRECTIVE

352   . Including a satellite

357.01	.. With position indicating
357.06	... Using Global Positioning Satellite (GPS or Glonass)
357.07	.... Tracking or monitoring (i.e., lost or stolen vehicles)
357.08	.... Determining relative position (e.g., distance or direction)
357.09	.... With transmission of location-indicative information to or from a remote station
357.1	.... Combined with telecommunication
357.11	.... Attitude determination
357.12	.... GPS receiver signal processing
357.13	.... With storage device (i.e., map or database)
357.14	.... Combined with secondary navigation system (i.e., LORAN, gyroscope, inertial, dead reckoning, etc.)
357.15	.... Satellite selection (i.e., tracking or acquisition)

Class 701 Data Processing: Vehicles, Navigation and Relative Location

200	NAVIGATION
207	. Employing position determining equipment
213	.. Using Global Positioning System (GPS)
214	... Means to improve accuracy of position or location
215	.... Having multiple GPS antennas or receivers (e.g., differential GPS)
216	.... Having an self-contained position computing means (e.g., dead reckoning)

Using the classification tools and browsing patents for these subclasses on the USPTO Web site helped to formulate search strategies to use in Dialog. It was found, for example, that many of the patents did not have GPS technology as a primary focus–that it represented only one aspect of the technology described in a patent record. One must decide how inclusively one wishes to define the technology under analysis. For legal patentability searches, the usual strategy is to be more comprehensive, and to rely on human judgement to narrow the search by browsing. For this type of analysis, however, one may wish to be more restrictive.

One strategy for focusing is to add keywords to your search. For example, the searcher could require that terms such as gps or global positioning be found in the title or abstract of these patents. However, this is often not a useful strategy for patents. Because patents represent new technologies, and technology from many countries, there may not be agreement on terms used. The earliest patents, in particular, may drop out of the set inappropriately. A more appropriate strategy to narrow a broad search is to require that the subject classification terms used are the primary class. The first class term listed on a patent is considered the primary classification, though many additional terms may be assigned. Some systems allow one to restrict to main classification as part of the search strategy–e.g., Dialog and LexisNexis–but for many systems this strategy requires browsing the records, whether online or offline.

This strategy can make a major difference in the number of patents retrieved, as Table 1 shows. Note that while column four (total) is the result of adding columns two and three (342 and 701, respectively), this is not true for the last column because relevant GPS subclasses from 342 and 701 may be assigned to the same patent. Requiring that one of the relevant GPS subclasses be assigned as the primary classification term resulted in a drop in the number of patents from 832 to 388 for the three-year period. It also appears that the class 701 terms are often assigned as secondary classifications rather than primary terms, given the major difference in the number of patents retrieved between classes 342 and 701 when primary class is used, in comparison to when this restriction is not used.

TABLE 1. Number of GPS Patents by Class

Year	Class 342 Primary	Class 701 Primary	Total–GPS Primary Class
2001	110	56	166
2000	95	22	117
1999	69	36	105
Total	274	114	388
Year	Class 342– Any Class	Class 701– Any Class	Total GPS Patents
2001	259	217	354
2000	170	183	271
1999	114	156	207
Total	543	556	832

To illustrate that GPS is a new technology, a distribution of the number of U.S. patents granted by year for a 20-year period follows in Table 2. These are patents that have been assigned at least one of the relevant classification terms, but not necessarily as a primary class. In the last three years, 1999-2001, 58.8% of all the patents have been granted. The previous three years (1996-1998) account for 25% of all the patents granted–less than half that of the most recent period. In the early years, a greater percentage of class 342 patents were granted than for class 701 (5% of the class 342 Communication patents in the first ten years, as compared to 2% of the class 701 Navigation patents; and 21.3% for the next five years for class 342, vs 16.6%)–that is, the satellite-related patents were granted prior to those for the application technologies related to navigation, though both have seen most of their activity in the past three years. One sees that GPS technology is rapidly emerging, and now rapidly evolving in all its subfields.

TABLE 2. Number of GPS Patents Granted, 1982-2001

Year	Class 342	Class 701	Class 342 and/or 701
2001	259 (27.3%)	217 (25.7%)	354 (46.3%)
2000	170 (17.9%)	183 (21.7%)	271 (19.2%)
1999	114 (12.0%)	156 (18.5%)	207 (14.6%)
1998	94 (9.9%)	98 (11.6%)	170 (12.0%)
1997	62 (6.5%)	30 (3.6%)	80 (5.7%)
1996	74 (7.8%)	47 (5.6%)	103 (7.3%)
1995	48 (5.1%)	42 (5.0%)	76 (5.4%)
1994	38 (4.0%)	26 (3.1%)	51 (3.6%)
1993	29 (3.1%)	14 (1.7%)	35 (2.5%)
1992	13 (1.4%)	11 (1.3%)	21 (1.5%)
1991	11 (1.2%)	3 (0.4%)	13 (0.9%)
1990	11 (1.2%)	11 (1.3%)	16 (1.1%)
1989	9 (0.9%)	2 (2.4%)	10 (0.7%)
1988	5 (0.5%)	0 (0)	6 (0.4%)
1987	5 (0.5%)	1 (0.1%)	5 (0.4%)
1982-86	6 (0.6%)	0 (0)	6 (0.4%)
TOTAL	**948**	**845**	**1414**

In this study, we are interested in mining the patent data to find out answers for the following questions:

1. Who are the major players in GPS technology development?
2. How diffuse is the GPS development–a few corporations or inventors in domination or many of them sharing the market?
3. How international is GPS technology development?
4. How active is the non-corporate involvement–are universities or government agencies strongly represented?

The data under analysis are the 388 U.S. patents granted in 1999-2001, that have the relevant GPS class terms assigned as primary classifications.

Identifying the Key Players

By the term 'key players,' we refer to both inventors and assignees–organizations to which the patents were granted. To explore the level of corporate R&D in developing GPS technologies for the U.S. market, its global nature, the 388 patents granted in 1999-2001 were first categorized by type of patentholder, and whether the patentholders were headquartered in the U.S. or other countries (Table 3). Many patent databases include the major categories developed by the USPTO–U.S. companies, U.S. government, foreign companies–as searchable features. These results were examined in order to make further distinctions, such as universities. For GPS patents granted in the U.S., technology development is primarily an American corporate activity, with 68% of the patents having U.S. corporate assignees. Few universities or government agencies hold patents–the very low rate is somewhat surprising when one considers that GPS began as a military effort, and much supportive policy-making regarding GPS has been enacted by the U.S. government.

A question for further research is whether any of the non-government patents still have links to the U.S. government. In the early 1980s, the patenting policy of the federal government changed with the implementation of the Bayh-Dole Act. Government-funded research conducted by other organizations that resulted in patents no longer required that the patents be held by the federal government. The organization conducting the research could be the patentholder, while the government retained an interest–usually involving the right to use the technology without paying royalties. In patent records, this is indicated by a field called 'statement of government interest,' which some U.S. patent databases make searchable. Further research using this field might tell us whether government funding is a factor in GPS technology development.

In addition to research outside the United States that was considered valuable for the U.S. market and therefore patented here, international collaboration appears in several other forms–patents are held by firms with U.S. headquarters, but inventors residing in other countries, or other combinations in which more than one country is involved, as Table 4 shows. An example is:

> US6232917. Navigational system.
> Inventors: Baumer, Jean-Claude (Cauges sur Mer Cedex, France); Giacalone, Jean-Claude (Vence, France); and Hilbig, Hans-Martin (Tiefenbach, Germany).
> Assignee: Texas Instruments Incorporated, Dallas, TX.

The geographic information for the assignee is based on the assignee's headquarters, whereas the inventor field shows the inventor's

TABLE 3. Distribution of GPS Patents by Type of Patentholder

Type of Patentholder	Number of Patents	Percentage of Patents
U.S. companies	264	68%
U.S. universities	5	1.3%
U.S. government agencies	6	1.5%
U.S. individual inventors	32	8.2%
Non-U.S. companies	71	18.3%
Non-U.S. government	2	0.5%
Non-U.S. universities	2	0.5%
Non-U.S. individual inventors	3	0.8%
Multiple assignees–U.S. and non-U.S. firms	3	0.8%
U.S. total	307	79.1%
Non-U.S. total	81	20.9%
TOTAL	**388**	**100%**

city and state (U.S.) or country (non-U.S.) of residence. Non-U.S. firms with U.S. subsidiaries may file patents for technology developed in the U.S. under the subsidiary's name and U.S. headquarters address; or multinational firms with branches in various countries may opt to file in the primary headquarters. Although 356 of the 388 patents list all inventors and assignees in the same country, 32 (8.2%) do involve more than one country.

Table 5 lays out this data on international collaboration from a slightly different perspective. If one looks at the countries for both assignees and inventors, how frequently does a patent list more than one country? The 32 patents with multiple countries represented include only three patents in which the United States is not the location of at least one assignee.

The next focus of analysis was specific key players. To determine the frequency of patenting activity by inventors and by assignees, the Dialog rank command was combined with reviewing the results. This feature should be used cautiously. Dialog treats as equivalent only exact matches–the searcher must look for those occurrences of different forms of the same organization or inventor name if one wishes to in-

TABLE 4. International Collaboration in GPS R&D

Type of Patentholder	Number of Patents	Percentage of Patents
U.S. organizations with U.S. inventors	254	65.5%
U.S. organizations with non-U.S. inventors	9	2.3%
U.S. organizations, with U.S. and non-U.S. inventors	12	3.1%
U.S. individual inventors	32	8.2%
Non-U.S. organizations, inventors same country	67	17.3%
Non-U.S. organizations, U.S. inventors	4	1%
Non-U.S. organizations, U.S. and non-U.S. inventors	1	0.3%
Non-U.S. organizations, non-U.S. inventors from 2 or more countries	3	0.8%
Non-U.S. individual inventors	3	0.8%
Multiple assignees, U.S. and non-U.S., inventors U.S. and non-U.S.	1	0.3%
Multiple assignees, U.S. and non-U.S., inventors non-U.S.	1	0.3%
Multiple assignees, U.S. and non-U.S., inventors U.S.	1	0.3%
TOTAL	**388**	**100%**

TABLE 5. International Collaboration in GPS R&D, by Number of Countries

Number	Number of Patents	Percentage of Patents
One country, U.S.	286	73.7%
One country, non-U.S.	70	18%
Two countries, U.S. and non-U.S.	26	6.7%
Two countries, both non-U.S.	2	0.5%
Three countries, including U.S.	3	0.8%
Three countries, all non-U.S.	1	0.3%
TOTAL	388	100%

crease the accuracy of the results. Two examples that require human judgement: Stanford University was listed as an assignee both in that form and as 'The Board of Trustees of the Leland Stanford Junior University.' Similarly, Nicholas C. Talbot and Nicholas Charles Talbot, both working for Trimble Navigation and filing patents with the same co-inventors, are likely the same person.

Notable Inventors

In ranking the inventors of the 388 patents, 22 were found to hold four or more patents. Norman F. Krasner, with 17 patents, was found to be the most active inventor in recent GPS technology development (see Table 6). The assignee for his patents is SnapTrack, which is the second largest holder of GPS patents, as will be shown later. Next is Mark E. Nichols, who holds eight patents, all with the assignee Trimble Navigation, the owner of the largest number of GPS patents (Table 9). Such information would be valuable for decision-making in personnel recruitment and other human resources activities, as well as merger and acquisition targeting, since corporations that have active outstanding inventors and developers would likely have an edge in R&D and other technology development. Table 6 lists the inventors that have at least five GPS patents, granted in 1999-2001.

Among the 388 GPS patents, many have multiple inventors (202, or 58.1%). Of the 562 people who have been granted GPS patents, 81% have one patent (Table 7). Only 3.9% are active recent inventors in this field, with four or more patents. The norm is the inventor with one patent.

TABLE 6. Inventors with Five or More GPS Patents

Number of Patents	Inventor Names	Assignee
17	Krasner, Norman F.	SnapTrack, U.S.
10	Nichols, Mark E.	Trimble Navigation, U.S.
8	Lin, Ching-Fang	American GNC is assignee for 2 patents; no assignee for 6 patents
7	Odagiri, Hiroshi	Seiko Instruments, Japan
7	Sakumoto, Kazumi	Seiko Instruments, Japan
6	King, Thomas Michael	Motorola is assignee for 5 patents; no assignee for 1 patent
5	Tsubata, Keisuke	Seiko Instruments, Japan
5	Eschenbach, Ralph	Trimble Navigation, U.S.

TABLE 7. Distribution of Activity Among Inventors

Number of Patents Per Inventor	Number of Inventors	Percentage of All Inventors
1	455	81%
2	71	12.6%
3	14	2.5%
4 or more	22	3.9%
Total Inventors	562	100%

Concentration of Technology Development

Many organizations have been involved in recent GPS technology development and patenting activities. As shown in Table 8, the 388 GPS patents are unevenly distributed among 142 assignees. The data demonstrates some degree of concentration; only a small number of organizations are highly active in the field. Most notable is Trimble Navigation, with 64 patents, accounting for 16.4% of the 388 patents (see Table 9). Of the nine firms holding seven or more patents, all but one are headquartered in the United States. The 112 organizations that were granted only one or two patents during 1999-2001 represent 78.9% of the 142 assignees, and their patents account for about 33% of the total. The nine most active firms hold 38% of the 388 GPS patents. These assignees are those that were granted the patents originally–those that

TABLE 8. Distribution of Activity Among Assignees

Number of Patents per Assignee	Number of Assignees	Percentage of All Assignees
1	91	64.1%
2	21	14.8%
3	11	7.7%
4 or more	19	13.4%
Total Assignees	**142**	**100%**

TABLE 9. Assignees with Five or More GPS Patents

Number of Patents	Percentage of All Patents	Assignee	Country
64	16.5%	Trimble Navigation	U.S.
19	4.9%	SnapTrack	U.S.
16	4.1%	Motorola	U.S.
13	3.4%	Rockwell Collins	U.S.
7	1.8%	IBM	U.S.
7	1.8%	Lucent Technologies	U.S.
7	1.8%	Qualcomm Inc.	U.S.
7	1.8%	Rockwell International	U.S.
7	1.8%	Seiko Instruments	Japan

conducted the research. Later mergers and acquisitions that resulted in patents changing hands are not reflected here.

An interpretation note: Holding the largest number of patents does not guarantee that the firm dominates the marketplace. Not all patents end up as products in the marketplace, and recent patents may not yet have had time to enter the market, since patents may be filed early in the product development process. As well, a firm may hold a small number of patents that are critical to the field and therefore give a strong market advantage. Additional information is needed to confirm market share. But the patent data does point to the likeliest competitors, and holders of recent patents may represent emerging competitors, who may not yet have many products in the marketplace.

The firms involved in GPS technology innovation are from many fields, in communications, computing, electronics and transportation. The widely distributed interests are consistent with the scenario that

GPS is just emerging. Emerging technologies typically have an early stage of highly dispersed development, followed by competition for market share and domination, during which mergers and acquisitions may become one of the major themes (Schmookler 1966, Trajtenberg 1990). If this type of patent analysis is done in order to identify potential candidates for merger or acquisition, then profiling the firms, using the typical company resources, is often the next step. Both of the top firms provide GPS-related technologies as their major products. SnapTrack was recently acquired by Qualcomm. (The patent records we used showed the assignee at time of issue; following acquisition, patents are usually reassigned to the new owner.)

Extent of Internationality

How international is GPS technology development? What countries have patented their GPS technology with the U.S. patent system? Because the U.S. is a large market, the U.S. patent system often reflects major international technical developments. Earlier discussion of types of patentholders found that 81 patents, or 20.9% of the 388, are held by organizations or individual inventors outside the United States, and three additional patents include both U.S. and non-U.S. firms as assignees. For the eight most active inventors, three are affiliated with Seiko Instruments, a Japanese firm, and Seiko is the one non-U.S. firm among the nine most active patentholders. Of the 388 patents, 97 (25%) include non-U.S. inventors. Thirty-one patents, or 8%, involve multinational activity, with inventors and/or assignees from more than one country (which may or may not include the U.S.).

The 388 patents include a total of 562 inventors if each individual is counted only once, but 764 inventors if one includes the same inventor counted multiple times, for each patent held. These 764 inventors are from 14 countries (see Table 10)–not a wide geographic variety. Japanese inventors are the most active. Inventors from Germany, Canada, Great Britain, Russia and New Zealand all received more than five GPS patents each in 1999-2001. The reasons for patenting their GPS technology in the U.S. may be many, among them: significant advances in infrastructure for GPS applications in these countries allow such developments and also urge such patenting in order to protect inventions; collaborative R&D activities across countries may prompt international patenting and facilitate doing so; and the U.S. is a major market one for GPS applications. Japan's high-tech products have been largely dependent on the markets of the U.S. and other major economic entities,

TABLE 10. Countries of Residence of GPS Inventors

Country	Number of Inventors	Percentage of All Inventors	Percentage of Non-U.S. Inventors
United States	568	74.3%	-
Japan	101	13.2%	51.5%
Germany	21	2.7%	10.7%
Canada	17	2.2%	8.7%
Great Britain	15	2%	7.7%
Russia	11	1.4%	5.6%
New Zealand	9	1.2%	4.6%
France	5	0.7%	2.6%
Israel	5	0.7%	2.6%
Australia	4	0.5%	2%
Sweden	4	0.5%	2%
Korea	2	0.3%	1%
Netherlands	1	0.1%	0.5%
Singapore	1	0.1%	0.5%
Total Inventors	764	100%	196 Non-U.S. Inventors

and their global positioning systems have relied heavily on the U.S. satellites and on their collaborations with such GPS key players as SnapTrack (Hofmann-Wellenhof et al. 1997, Steede-Terry 2000). For U.S. firms, countries such as Japan, New Zealand, Canada, and Great Britain are both potential competitors and collaborative partners in GPS technology developments.

FURTHER RESEARCH

Although we can learn much with the techniques illustrated here, additional technical assessment research is possible. One might compare the USPTO patent class schedule and the International Patent Classification (IPC), and explore the disciplinary distribution of GPS patents. International patent data can be used to look at families of patents. When an applicant files on the same invention in multiple countries, these applications and the subsequent publications are collectively known as a patent family. By looking at the patent family, one can get the global picture of where the company or individual has chosen to seek protection for its invention. This helps to track the patent holder's geographic focus for marketing its potential products. The size of the family sometimes can also tell a great deal about the perceived significance of the invention. The USPTO patent database alone cannot support such a task; one must also work with an international patent database. Using GIS and statistical software to map and model the spatial pattern and family chain of such patents could be very compelling, because graphical and spatial presentation would help us visualize the data and its statistical and geo-spatial patterns, enhance our understanding of the data, and crystallize valuable information from the data.

Another question about geographic distribution arises. Is there any specific pattern to the spatial distribution of GPS patents within the U.S. or, in other words, how are the patents distributed across different states or regions in this country? One might explore whether the spatial concentration of GPS patents in the U.S. confirms Feldman's (1994) argument that "innovation is expected to exhibit strong geographic clustering because new product commercialization relied on knowledge that is cumulative and place-specific" (Feldman, 1994: 29). According to Feldman, there is evidence that geographic clustering has been important to technological advancement throughout history. The Silicon Valley phenomenon in California provides a prominent example of this type of geographic clustering. The attraction for innovative firms to agglomeration of needed resources is so strong that it is likened to a magnet (Florida and Kenney 1990: 54-55).

Other issues for future research include a data analysis based on application dates rather than patent issuing dates, a comparison of patent data patterns between different subclasses, and a more detailed longitudinal and/or time series analysis. Since there is usually a one- to three-year time lapse between the filing date and granting date for a patent application, with the range varying widely, data based on application dates may help to track related technology developments. With a comparison of patent data patterns across subclasses, we could know more about the types and content of innovations that have come to a class or a field of technology. Longitudinal and time series analysis would facilitate finding out long-term development trends.

Frequently cited patents have also been found to be technically valuable (Hall et al. 2000). To learn more about knowledge diffusion in the GPS field (Jaffe et al. 2000), and to identify which patents are more technically important, citation analysis would be a must in future research.

CONCLUSIONS

This study has assessed certain aspects of the recent innovations and development of GPS technology as represented in U.S. patents. It has demonstrated the value of patent data for technology assessment. In the analysis, GPS technology is found to be a rapidly emerging and changing field, dominated by a few American corporations and geographically concentrated, but of interest to many organizations in different fields and a number of countries. The findings show that patent data could provide a sound foundation for evaluating technological innovations and R&D development trends. In spite of some shortcomings, as Griliches (1990) contends, "Patents statistics remain a unique resource for the analysis of technical change. Nothing else even comes close in the quantity of available data, accessibility, and the potential industrial, organizational, and technological detail."

REFERENCES

Albert, Michael B. and D. Hicks. 1998. Using Patent Data for Strategic Planning. *SATM Innovation & Technology Management News* 2 (2).

Delphion, Inc. *U.S. Granted Patents.* Available at: http://www.delphion.com/. Accessed November 2002.

Derwent World Patents Index. Thomson Derwent. Available at: http://www.derwent.com/worldpatentsindex/index.html. Accessed November 2002.

Feldman, Maryann P. 1994. *The Geography of Innovation.* Boston, MA: Kluwer Academic Publishers.

Florida, Richard and Martin Kenney. 1990. *The Breakthrough Illusion: Corporate America's Failure to Move from Innovation to Mass Production.* New York: Basic Books.

GARMIN International. *About GPS.* Available: http://www.garmin.com/aboutGPS/. Accessed November 2002.

Gregory, James and K. Mulligan. 1979. *The Patent Book: An Illustrated Guide and History for Inventors, Designers, and Dreamers.* New York: A & W Publishers.

Griliches, Zvi. 1984. *R&D, Patents and Productivity.* Chicago, University of Chicago Press.

_____. 1990. Patents: Recent Trends and Puzzles. *Brookings Papers on Economic Activity, Microeconomics*: 291-330.

Hall, Bronwyn, Adam B. Jaffe and Manuel Trajtenberg. 2000. *Market Value and Patent Citations: A First Look.* NBER Working Paper no. 7741, National Bureau of Economic Research, Cambridge, MA.

Hofmann-Wellenhof, B. et al. 1997. *Global Positional System: Theory and Practice.* New York: Springer-Verlag Wien New York.

Jaffe, Adam B., Manuel Trajtenberg and M. S. Forgarty. 2000. Knowledge Spillovers and Patent Citations. *American Economic Review* 90 (2): 215-218.

Kehoe, Cynthia A. 2001a. *Major Players in the International Rose Market: Using Patent Analysis to Examine an Industry*. Working Paper ISRN UIUCLIS–2001/3+INFO Graduate School of Library and Information Science, University of Illinois at Urbana-Champaign.

_____. 2001b. *Patents for Business Intelligence*. Presentation, National Technological University, February 2001.

Kehoe, Cynthia A. and Xiao J. Yu. 2001. Patent Data for Technology Assessment, Part I: Applications, Patent Databases, and Retrieval Methods. *Science & Technology Libraries* 22 (1/2): 101-116.

Mogee, Mary Ellen. 1997. Patents and Technology Intelligence. In: *Technical Intelligence for Business: Keeping Abreast of Science and Technology*. Edited by W. B. Ashton and R. A. Klavans. Washington, DC: Battelle Press.

Steede-Terry, Karen. 2000. *Integrating GIS and the Global Positional System*. Redlands, CA: ESRI Press.

Trajtenberg, Manuel. 1990. *Economic Analysis of Product Innovation: The Case of CT Scanners*. Cambridge, MA: Harvard University Press.

Trimble Navigation Limited. *All about GPS*. Available: http://www.trimble.com/gps/. Accessed November 2002.

U.S. Patent and Trademark Office (USPTO). *Classification Definitions*. Washington, DC: U.S. Government Printing Office. Also available at: http://www.uspto.gov/web/patents/classification/. Accessed November 2002.

_____. *Index to the United States Patent Classification System, December 2001*. Washington, DC: U.S. Government Printing Office, 2002. Also available at: http://www.uspto.gov/go/classification/uspcindex/indextouspc.htm. Accessed November 2002.

_____. *Manual of Classification*. Washington, DC: U.S. Government Printing Office. Also available at: http://www.uspto.gov/go/classification/. Accessed November 2002.

_____. *Patent Full-Text and Full-Page Image Databases*. Available at: http://www.uspto.gov/patft/. Accessed November 2002.

_____. *Products and Services Catalog 2001*. Available at: *http://www.uspto.gov/web/offices/ac/ido/oeip/catalog/index.html*. Accessed November 2002.

U.S. Patents Fulltext. Dialog Corp. Available at: http://library.dialog.com/bluesheets/html/bl0654.html. Accessed November 2002.

USPAT, *U.S. Utility, Design and Utility Patents*. LexisNexis. Available at: http://web.nexis.com/sources/scripts/info.pl?228804. Accessed November 2002.

Vijay-Rao, Geeth. 2001. Creating Intelligence from Data: Resources and Techniques for Leveraging Patent Information for Competitive Advantage. In: *National Online Meeting Proceedings 2001*, 505-534. Edited by Martha E. Williams. NY: Information Today.

TRADEMARKS

Finding Your Way Through the Trademark Information Maze

Charlotte A. Erdmann

SUMMARY. This article gives guidance in searching trademarks, owner names, goods and services, and international classes using sample queries in the trademark databases of the United States Patent and Trademark Office (USPTO). Web and DVD databases are described. Both systems have advantages and disadvantages. The author presents an introduction to trademark terminology and the basic application process. Five queries are completed showing interesting uses of trademarks as well as the strengths and weaknesses of the databases. *[Article copies available for a fee from The Haworth Document Delivery Service: 1-800-HAWORTH. E-mail address: <docdelivery@haworthpress.com> Website: <http://www.HaworthPress.com> © 2001 by The Haworth Press, Inc. All rights reserved.]*

Charlotte A. Erdmann is Assistant Engineering Librarian and Associate Professor of Library Science, Siegesmund Engineering Library, Potter Center, Purdue University, 500 Central Drive, West Lafayette, IN 47907-2022 (E-mail: erdmann@purdue.edu). She was a Fellowship Librarian at the United States Patent and Trademark Office in 1998-1999.

[Haworth co-indexing entry note]: "Finding Your Way Through the Trademark Information Maze." Erdmann, Charlotte A. Co-published simultaneously in *Science & Technology Libraries* (The Haworth Information Press, an imprint of The Haworth Press, Inc.) Vol. 22, No. 1/2, 2001, pp. 137-160; and: *Patent and Trademark Information: Uses and Perspectives* (ed: Virginia Baldwin) The Haworth Information Press, an imprint of The Haworth Press, Inc., 2001, pp. 137-160. Single or multiple copies of this article are available for a fee from The Haworth Document Delivery Service [1-800-HAWORTH, 9:00 a.m. - 5:00 p.m. (EST). E-mail address: docdelivery@haworthpress.com].

http://www.haworthpress.com/store/product.asp?sku=J122
© 2001 by The Haworth Press, Inc. All rights reserved.
10.1300/J122v22n01_09

KEYWORDS. Trademarks, service marks, certification marks, collective membership marks, *Cassis2*, *Trademark Electronic Search System* (*TESS*)

INTRODUCTION

The goal of this article is to give guidance in searching trademarks and owner names by using sample queries in the trademark databases of the United States Patent and Trademark Office (USPTO). The USPTO publishes two trademark database systems: *Trademark Electronic Search System (TESS)*, a Web-based database updated daily, and *Cassis2*,[1,2] a DVD product updated bimonthly. A third database called *X-Search* is available to trademark examining attorneys and the Detroit and Sunnyvale Patent and Trademark Depository Libraries. *X-Search* is not covered by this article since it is available at a limited number of sites. This article does not give detailed descriptions of database display screens and does not explain all features of the systems. The reader is encouraged to review database help screens and training materials that accompany *Cassis2* and *TESS*.

This article gives an introduction to trademarks but makes no attempt to explain the complete trademark process or fees involved in filing applications. The basic process is discussed in two publications, *Basic Facts about Trademarks*[3] and *Frequently Asked Questions about Trademarks*.[4] The *Trademark Manual of Examining Procedure* (TMEP)[5] is the examining attorney's manual and gives detailed explanations of terminology, laws and regulations, and procedures.

Nothing in this article should be considered legal advice. Applicants, registrants, and those affected by marks should consult legal counsel.

DEFINITIONS AND BASIC INFORMATION

The USPTO[6] defines a mark as "a word, phrase, symbol or design, or a combination of words, phrases, symbols or designs" that are "used, or intended to be used, in commerce." A federal registration[7] may be given to a mark used on goods or services in interstate commerce or commerce between the U.S. and another country. The goods must cross state lines and the mark must be displayed on the goods or its packaging. With services, this "involves offering a service to those in another state or rendering a service which affects interstate commerce (e.g., restaurants, gas stations, hotels, etc.)."[8] A federal registration gives the owner of the mark rights throughout the United States and its territories.

One may apply for a state registration when the goods or services are used within a state. The Office of Secretary of State is responsible for the registration of state trademarks in most states. In several states, another office handles the registration. Michael White[9] has compiled a Web site for *State Trademarks* with links or addresses to each state office. Another option is common law trademarks.

Kevin Harwell has written an excellent article on "Resources for Searching Common Law Trademarks."[10] Harwell states, ". . . the trademark owner may elect not to register the trademark at all. Federal law and the laws of some states provide some protection for such marks, known as common law marks. Protection for registered trademarks is generally regarded as much better than common law trademarks." He recommends searching a variety of resources, including online and print sources.

Federal applications may be filed based on "use in commerce" or "intent to use." The USPTO[11] refers to "use in commerce" as "bona fide use of the mark in the ordinary course of trade." A "use in commerce" application cannot be filed to "reserve rights to a mark." The trademark must appear on the goods, its containers, or displays associated with the goods. The service mark "must be used or displayed in the sale or advertising of services." "Intent to use" applications are filed when one has not used the mark, but "plans to do so in the future." The application may be filed "based on a good faith or bona fide intention to use the mark in commerce."

The author includes definitions detailed by the USPTO[12] for trademark, service mark, certification mark, or collective mark (Table 1). These marks are eligible for federal registration.

Goods and services must be specified in the application. The USPTO recommends that applicants use the *Acceptable Identification of Goods and Services Manual*[13] to choose the terminology used to describe their goods or services. The Manual also lists the International Class for each good or service. There are 45 International Classes, 34 for Goods and 11 for Services, based on the *Nice Agreement Concerning the International Classification of Goods and Services.*[14] Filing fees are charged per class of goods/services. The trademark examining attorney assigns a class or classes if the applicant does not. Applicants may designate as many classes as they wish in a single application, but currently the fee for each class is $335.

An application may be denied when one files an application that is confusingly similar to another mark. This is particularly true when

TABLE 1. Definitions

Trademark	Identifies and distinguishes the "goods of one manufacturer or seller from goods manufactured or sold by others, and to indicate the source of the goods. In short, a trademark is a brand name." The stylized form of Coca-Cola[1] has been used in commerce for nutrient or tonic beverages since 1887. It was registered in 1892.
Service Mark	Identifies and distinguishes the "services of one provider from services provided by others, and to indicate the source of the services." ENRON Corp. has a variety of pending and registered service marks for ENRON.
Certification Mark	Is used with the "owner's permission by someone other than its owner, to certify regional or other geographic origin, material, mode of manufacture, quality, accuracy, or other characteristics of someone's goods or services, or that the work or labor on the goods or services was performed by members of a union or other organization." The use of ⓤ[2] on a product certifies that those who use this mark have done representative samplings of goods to conform to the requirements of Underwriters Laboratories.
Collective Membership Mark	Is used by "members of a cooperative, an association, or other collective group or organization." Use of the mark "indicates membership in a union, an association, or other organization." Navajo Code Talkers[3] is an example of a collective membership mark.

[1] U.S. Trademark Registration 0022406. *Cola-Cola [Words, letters, and/or numbers in a stylized form]*. Coca Cola Company. Dated January 13, 1893; 4th Renewal, January 13, 1983. Expires January 13, 2003.
[2] U.S. Trademark Registration 0782589. ⓤ. Underwriters Laboratories, Inc. Dated December 29, 1964, 1st renewal, March 26, 1985. Expires March 26, 2005.
[3] U.S. Trademark Registration 2487105. *Navajo Code Talkers*. Navajo Code Talkers Association. Dated September 11, 2001. Expires September 11, 2011.

goods or services are similar. Phyllis Karrh and Robin Kelley[15] have written an exceptional article, "Trademarks: More Than Meets the Eye," that describes why and how one should file a federal application for trademark protection vs. state registration. The article also highlights the principles of determining if a mark is confusingly similar to another. Some additional reasons for rejections are briefly discussed in "Can the Office refuse to register a mark?"[16]

An application is pending until it has been examined by a trademark examining attorney and meets all the requirements for registration. At that time, the application is published for opposition in the print *Official Gazette of the United States Patent and Trademark Office: Trademarks*[17] and in the online *Official Gazette–Trademarks (OG)*[18] (most recent five weeks available). If there is no opposition, the mark usually registers within twelve weeks. If there is opposition, the mark is reviewed by the Trademark Trial and Appeal Board (TTAB). Decisions of the TTAB are summarized in the online *Official Gazette* as well as the print *OG*. Information about pending, registered, and dead marks is open to the public. The databases are described later in this article.

When registered, those applications for trademarks already in use receive a registration certificate. Intent-to-use applications receive a notice of allowance. A statement of use must be filed before the applicant receives a registration certificate. Once registered, the mark must continue to be used in commerce in order to be "active." Forms are filed and fees are paid to show continued use and to receive renewals. Marks that are no longer used or abandoned are called "dead."

According to the *TMEP*,[19] "an application to register a mark must be filed by the owner of the mark or, in the case of an intent-to-use application under 15 U.S.C. §1051(b), by the person who is entitled to use the mark in commerce. Normally the owner of a mark is the person who applies the trademark to goods produced by him or uses the service mark in the sale or advertising of services performed by him."

The owner may be a person, corporation, partnership, joint venture, union, and other organization that can sue another in court. "Nations, states, municipalities, and other related types of bodies operating with governmental authorization may apply to register marks that they own."[20]

DATABASES

Databases are published on the USPTO Web site and *Cassis2* DVDs. The most important information for searching pending, active, and recent dead marks is available on both systems. The Web databases and *Cassis2* software both have sophisticated search systems. Some information is only on one system or the other. Both systems have advantages and disadvantages. *DocDW* software is used for viewing the Trademark Registrations on the *Cassis2* workstations. Access is only available by Registration Number. Table 2 compares data in the USPTO Web databases and *Cassis2/DocDW* databases.

The *Trademark Electronic Search System (TESS)* on the USPTO's Web site contains over 3.2 million pending, dead, and active marks. The introduction[21] to the *TESS* Help indicates that the "database does include the available information on inactive applications and registrations (i.e., abandoned applications or canceled or expired registrations). Information on applications and registrations that were inactive prior to 1984, however, are generally not available on *TESS*." For up-to-date information about recently added applications, one should consult *Trademark Electronic Search System News*.[22] The dates given may not be

TABLE 2. Comparison of Web Databases and *Cassis2/DocDW* Products

Data Type	USTPO Web	*Cassis2/DocDW*
Basic Information, Guides, Manuals, Laws & Regulations	Trademark Guides, Trademark Manuals, Laws & Regulations	*Trademarks ASSIST*
Pending, Active, and Dead Marks	*Trademark Electronic Search System (TESS)*	*Trademarks BIB*
Status or History of Application	*Trademark Applications and Registrations Retrieval (TARR)*. Links available from *TESS* to *TARR*	Not Available
Trademark Assignments	Limited data available on *TESS* (some out-of-date)	*Patents and Trademarks ASSIGN*
Trademarks Published for Opposition	*TESS*. Also available in most recent five on-line issues, *Official Gazette–Trademarks*	*Trademarks BIB*
Full-Text Trademark Registrations	Most recent five on-line issues, *Official Gazette–Trademarks*	*USAMark* on CD-ROM
Trademark Trial and Appeal Board, Status of Proceedings	*The Board Information Systems Index (BISX)*	Not Available

complete. No average delay data is available. Based on data from the *News*, one assumes that electronic applications, i.e., e-TEAS, are added to *TESS* faster than paper applications. The *Trademark Applications and Registrations Retrieval (TARR)* system shows the status and history of the application. Users are encouraged to read the limitations[23] on currency and content of TARR so that one may understand what is and is not in the database. There is a link between the *TESS* and *TARR* databases. The *Board Information Systems Index (BISX)* shows the status of proceedings before the Trademark Trial and Appeal Board. The author is preparing a separate article on tracking proceedings of the TTAB.

USPTO staff members also create three *Cassis2* trademark databases on DVD. *Cassis2* software has three search methods (short form search, form search, and command search) and allows sophisticated searching, printing, and exporting. The databases include *Trademarks BIB*, *Patents and Trademarks ASSIGN*, and *Trademarks ASSIST*. *BIB* and *ASSIGN* are published every two months while *ASSIST* is published irregularly. The USPTO also produces *USAMark* monthly on CD. *USAMark* contains registration certificates and images and uses document delivery software called *DocDW*. These registration certificates are also available for the most recent five weeks of the online *OG-Trademarks*.

Trademarks BIB contains bibliographic information about active registered trademarks and pending trademarks as well as trademarks that have been abandoned, cancelled, and expired since 1984. Some dead marks before 1984 are available. *Trademarks BIB* is a text database and does not contain images of trademarks that are design only or designs in combination with text. Design codes are available for searching. *Patents and Trademarks ASSIGN* includes data for both patents and trademarks. Patent data is derived from assignment deeds recorded after August 1980 and trademarks recorded at the USPTO from 1984-date. Some earlier trademark assignment information (i.e., 1955-1983) is contained in *ASSIGN* but it is very incomplete. Patents and trademarks may be searched at the same time or separately. The *Trademarks ASSIST* includes searchable text of many trademark manuals. It contains the *Goods and Services Manual, Trademark Manual of Examining Procedure*, the *Trademark Trial and Appeal Board Manual of Procedure*, the *Trademark Statute and Rules (Trademark Act of 1946 and the Rules of Practice)*, examination notes and telephone index.

The *USAMark* CD-ROM contains images of the original printed documents of U.S. registered trademarks from 1870 to the present. The documents are retrieved by registration number only. *USAMark* indicates that a registration was granted at one time. It does not indicate the status of the mark. The registration may be active, abandoned, cancelled, or expired. One may check *TESS* and *Trademarks BIB* for status of marks since 1984. It is possible to import registration numbers identified on the *Trademarks BIB* database into *USAMark*. One may also import numbers from any source in ASCII text format.

Table 3 provides a comparison of availability and system features and shows some of the advantages and disadvantages of the USPTO Web databases and the *Cassis2* and *DocDW* software systems.

Heavy users of trademarks may want to buy data on cartridge, CD-ROM, DVD-ROM or FTP from the USPTO. One should consult the *Products and Services On-Line Catalog*[24] to obtain a list of trademark products. Commercial information hosts that specialize in trademark databases are also recommended.

TWO QUICK SEARCHES

An individual may complete a trademark search for a variety of reasons. These may include: curiosity, new application filings, potential infringement cases, competitive intelligence, history, job interviews, or possible licensing opportunities.

TABLE 3. Availability and System Features

Advantages	Web Databases	Cassis2 and DocDW Databases
Users of Databases	Individuals, libraries, and businesses with Web access.	Subscribers, clients of Patent and Trademark Depository Libraries, and some GPO Depository Libraries.
Hours Available	23 hours a day, updated from 4:00 a.m. to 5:00 a.m., Tuesday through Saturday.	Business hours
Browse or Index features	Limited browse capability.	Sophisticated browse capability using Show Index Icon.
Ease of Printing and Exporting/Importing	Use of Web browser is necessary to print and save list and/or the complete record.	Sophisticated printing and exporting/importing capabilities.

A curious person may want to discover who owns a specific mark using *TESS* or *Trademarks BIB*. To search for the mark, Kevlar, a search is done in Combined Word Mark field to retrieve terms from the Basic Index, Translation Index, Mark Punctuated (Word Mark), and Translation Statement. Table 4 shows one of several registrations owned by Du Pont for Kevlar.

When a searcher is focusing on marks that have symbols or designs, it is best to examine Design Search Codes. These identify design elements that comprise a mark, e.g., a bear. The codes are outlined in the *Design Search Code Manual*[25] and are assigned by Trademark Office staff. An application may have more than one code. Guidelines and an alphabetical index make the manual easy to use and understand. A design for a bear is part of design code 03.01 for "Cats, dogs, wolves, foxes, bears." Table 5 shows the relevant codes for bears in more detail.

When one looks at Table 5, it shows the design codes: 03.01.14, 03.01.24, and 03.01.26 that are used on a registration for a bear whose owner is Grateful Dead Productions. Table 6 displays the registration.

An applicant is required to identity Goods and Services (G & S) in the application, e.g., "clothing; namely, T-shirts, sweatshirts, jackets, hats, caps and socks." These Goods are part of International Class 025 that involves: "Clothing, footwear, headgear."[26] A searcher may want to consult the *Acceptable Identification of Goods and Services Manual*[27] to choose the Goods and Services terminology. The manual also lists the classes.

TABLE 4. One of the Registrations for Kevlar (Record from *TESS*)

Word Mark	KEVLAR
Goods and Services	IC 022. US 001. G & S: MAN-MADE FIBERS FOR GENERALIZED USE IN THE INDUSTRIAL ARTS. FIRST USE: 19730406. FIRST USE IN COMMERCE: 19730406
Mark Drawing Code	(1) TYPED DRAWING
Serial Number	72455205
Filing Date	April 20, 1973
Registration Number	0983080
Registration Date	May 7, 1974
Owner	(REGISTRANT) E. I. DU PONT DE NEMOURS AND COMPANY CORPORATION DELAWARE 1007 MARKET ST. WILMINGTON DELAWARE 19898
Attorney of Record	ARNETTA MCRAE
Type of Mark	TRADEMARK
Register	PRINCIPAL
Affidavit Text	SECT 15. SECT 8 (6-YR).
Renewal	1ST RENEWAL 19940627
Live/Dead Indicator	LIVE

FIVE QUERIES

A variety of queries can be answered by searching owner information. Major fields that are also used include: Marks, Goods and Services, and International Classes. Some queries are straightforward and others are more complicated. It is important to determine fields that may contain the needed information. Guidance is provided in the database help screens as well as any available manuals. Assistance with field searching, truncation, Boolean operators, and proximity operators should also be reviewed.

The author has created five queries that are examples of fictitious searches. The scenarios are discussed in the next section:

- A competitive intelligence professional who works for Frost & Sullivan is working on a project to analyze companies that currently sell pacemakers or are planning to begin sales.
- A history student knows that the first trademark was registered in 1870. He wants to identify the oldest active registered U.S. trademark and other trademarks by the related owner(s).

- XYZ Police Vests, a fictitious applicant, is filing a mark for bullet-proof vests and wants to avoid choosing a mark that might be confused with a current mark. XYZ wants to see the broad picture of marks that are currently used so that their company can choose a unique mark, save marketing costs, and avoid future litigation.
- The eBay Trading Post Company, which runs an online auction and flea market service, is concerned that other companies may be attempting to register marks that might be confusingly similar to its own.
- An independent inventor is looking for companies that might license his patented invention for an integrated television and Internet system. The person does not have the financial resources or technical expertise to bring this product to market. Someone interested in this query may typically search patent databases to identify licensing opportunities.

OWNER TERMINOLOGY

The easiest way to answer the queries may be to search the Owner Name in combination with other fields. Sometimes one may search the Goods and Services field and analyze results including the owner.

Owner Name is the preferred field in *TESS* and *Trademarks BIB*. One can expect that names change over time. Owners are called an "Applicant" or "Registrant" in the output displays of both databases depending on whether an application is pending or registered.

Names of individual owners are inconsistently entered. Some have first name middle name/initial last name. Others are entered last name first name middle name/initial. Jamie Lee Curtis is listed as "Jamie Lee

TABLE 5. *Design Search Code Manual*: Bears

03.01.13	Panda bears
03.01.14	Other bears
03.01.16	Heads of animals in this division
03.01.17	Animals in this division with forepaws resting on a shield, crest or other
03.01.24	Stylized animals in this division
03.01.26	Costumed animals and those with human attributes in this division

TABLE 6. Registration for Dancing Bear (Record from *TESS*)

Goods and Services	IC 025. US 039. G & S: clothing; namely, T-shirts, sweatshirts, jackets, hats, caps and socks. FIRST USE: 19890601. FIRST USE IN COMMERCE: 19890601
Mark Drawing Code	(2) DESIGN ONLY
Design Search Code	030114 030124 030126
Serial Number	74142311
Filing Date	February 25, 1991
Published for Opposition	June 30, 1992
Registration Number	1718272
Registration Date	September 22, 1992
Owner	(REGISTRANT) Grateful Dead Merchandising CORPORATION CALIFORNIA P.O. Box 12979 San Rafael CALIFORNIA 94915
Assignment Recorded	ASSIGNMENT RECORDED
Attorney of Record	MICHAEL J DALTON
Type of Mark	TRADEMARK
Register	PRINCIPAL
Affidavit Text	SECT 15. SECT 8 (6-YR).
Live/Dead Indicator	LIVE

Curtis." Frank Sinatra is listed as "Sinatra Frank." Both have active marks. When an individual is listed as Jr. or III, this is also inconsistently entered in the database. For example, Dennis C. Redden is listed as "Dennis C. Redden III" and "Redden, Dennis C." A corporation may have subsidiaries, change its name, or have more than one name in the database. For example, Harley Davidson, Inc. and H-D Michigan, Inc. are both used. There are also other motorcycle-related owned assignments for Harley Davidson, including local sales dealerships.

USPTO staff report that some data has been converted to machine readable form by OCR which may account for some of the inconsistent entry. Applicants may enter their names inconsistently on applications. Examining attorneys may correct some errors on pending applications.

In *Patents and Trademarks ASSIGN*, owners are searched using the terms, Assignor and Assignee. Assignment means a transfer by an owner of "its entire right, title and interest in a registered mark or a mark for which an application to register has been filed."[28] When the assignment deed is recorded at the USPTO, there is a record in the *ASSIGN* database. For example, a company from Finland assigned the mark Kevlar to Du Pont. The Assignor transfers ownership to the Assignee. Many times a single assignment record has multiple trademark transfers. When part of the Hewlett-Packard Company became Agilent Technologies, Hewlett-Packard (Assignor) transferred ownership of many trademarks to Agilent Technologies (Assignee).

SAMPLE QUERIES

Queries and suggested strategies are discussed here for information only and do not attempt to show every conceivable way that a query may be answered. The strategies are intended to help the reader understand the databases and encourage creative use of them. All strategies may be entered in upper or lower case. Many sample queries require the use of the *Trademark Acceptable Identification Goods and Services Manual*[29] for choosing Goods and Services and International Classes. *Basic Facts About Trademarks* contains a list of International Classes for Goods and Services and may also be consulted.[30]

TESS has three search mechanisms: new user search, structured search, and free form search. To do complicated searches, free form is recommended. Fields may be qualified with [tag] or .tag. Help screens show [tag] so all examples are shown this way. On *TESS*, heavy use of truncation may result in "word overflow limit" messages. *TESS* Help includes a Frequently Asked Question[31] on the topic and explains that the message "is usually the result of a complicated search that mixes truncation with Boolean or proximity operators." The FAQ also offers advice on how to solve the problem. *TESS* uses the Web browser for displaying and printing records one at a time. It is very tedious with a lot of records.

Cassis2 databases, including *Trademarks BIB* and *Patents and Trademarks ASSIGN*, have three search modes: short form search, form search, and command search. The default search for short form search and form search is AND while the command searches uses OR. The form search and command search have more features. In command search, fields may be qualified with .tag. *Cassis2* software has advantages for sorting, marking, printing, and exporting records. Print and export outputs can be customized to specific fields. Data can be later imported into a database management or spreadsheet program for further analysis. Registration numbers may be exported from *Trademarks BIB* and imported into *USAMark*. The registration numbers can be imported from an ASCII text file.

Query 1

A competitive intelligence professional who works for Frost & Sullivan is working on a project to analyze companies that currently sell pacemakers or are planning to begin sales. Trademarks may be one of the sources that the competitive intelligence professional (CI) consults. The CI is looking for owners of trademark applications and registrations for pacemakers. A job seeker looking for employment in engineering design or marketing may also do a similar search. The Goods and Services field is used for this search. One considers all synonyms and similar heart related products. The following terms may be considered:

>pacemaker
>defibrillator
>(heart or cardiac or cardiology) and (rhythm or rate or regulation)

Truncation symbols are added as needed. See the online help for a complete listing of symbols.

The *Trademark Acceptable Identification Goods and Services Manual* shows that the preferred term is heart pacemakers in International Class 010. This class is for "Surgical, medical, dental, and veterinary apparatus and instruments, artificial limbs, eyes, and teeth; orthopedic articles; suture materials."

Searches may be done on either system, *TESS* or *Cassis2 Trademarks BIB*. Both systems offer online help to explain operators, truncation symbols, and special features.

>a. In *TESS*, one may search:

>>heart adj pacemakers[gs]   results in 169 records
>>pacemak$[gs]   results in over 500 records

>When one tries a more thorough strategy, a "word limit overflow" message halts this search:

>>(pacemak$ or defibril$ or ((rhythm$ or rate$ or regulation) and (heart$ or cardiac or cardiol$)))[gs]

By eliminating some truncation, this search results in 1,353 records:

((pacemak$ or defibril$ or ((rhythm or rate or rates or regulation) same (heart or hearts or cardiac or cardiology))))[gs]

Some marks are not relevant. When one adds the International Class: "010"[ic], it results in over 1,076 marks.

b. In *Trademarks BIB*, this strategy works best in the command search. The database is updated every two months. Information is slightly dated. This strategy retrieves 1,353 marks:

((pacemak$ or defibril$ or ((rhythm$ or rate$ or regulation) and (heart$ or cardiac or cardiol$)))).gs.

Adding the International Class, "010".ic. results in 1,051 records.

In *Cassis2* databases, display styles may be customized for viewing, printing, and exporting records. Sorting is also available but is slow for a large number of records. It may be better to export the records to database management or spreadsheet programs and sort using one of these programs. In this case, sorting by owner name is very helpful.

Query 2

A history student knows that the first trademark was registered in 1870. He wants to identify the oldest active registered U.S. trademark and other trademarks by the related owner(s). The student thinks that there may be active registrations between 1870 and 1890.

A registration may be renewed as long as the trademark continues to be used in commerce. Related owners may involve a number of scenarios. Owner names sometimes change with the renewal. As a result, several owner names are searched to give a complete picture of the ownership history. Ownership may also be transferred by assignment.

To complete this query, four fields are useful: Registration Number, Registration Date, Filing date, and Live/Dead. Definitions obtained from *Trademark Electronic Search System Help: Search Fields*[32] are listed in Table 7.

In *TESS*, the Browse Dictionary feature does not accommodate a query on registration date. PTO staff indicates that in theory one should be able to browse any field. In practice, some fields fail during heavy use periods and cause timeouts for the Web browser before a replay is provided by the search engine. When the browse feature works well, it

TABLE 7. Field Definitions for Query 2

Field	Definition	Field Tag	Sample Search
Registration Number	Unique number assigned to applications that have received approval for registration. The registration number must be higher than 0. Incomplete and pending applications are given the number 0.	[RN]	To find registrations larger than 0, type: `rd > 0
Registration Date	Date on which a mark was registered by the US Patent and Trademark Office.	[RD]	To find a registration date before 1890, type this query: `rd < 18900000
Filing Date	Date when a complete application was received by the US Patent and Trademark Office, following receipt of all filing material requirements. Using this field may find applications that are pending.	[FD]	To find a filing date before 1890, type this query: `fd < 18900000
Live/Dead	LIVE[LD] can be used to obtain all live records. Live may mean that it is registered or pending. DEAD[LD] means that the application is no longer under prosecution within the USPTO.	[LD]	To find only live applications, type: live[ld]
Owner Name	Name of the individual, corporation, partnership, association, etc., having controlling interest in the use of the mark.	[ON]	(samson adj ocean adj systems)[on] or (tolman near1 jampes) [on]

can be a wonderful feature. When it does not work, it ties up a lot of time before the user gets frustrated and tries another search method. Here is the method used:

a. Since the first trademark law passed in 1870, one can assume that the earliest active registration date may be prior to 1890. There are many pending and incomplete records that show the registration number is 0. As a result when a registration number is larger than 0 and live, the trademark is active. Help screens suggest that this search may answer the question:

(`rd < 18900000) and (`rn > "0") and live[ld]

This results in seven trademarks. The author notices that there is one irrelevant registration that has no registration date. Jumping to the last mark, 0011210,[33] one retrieves *Samson* (a design plus words, letters, and/or numbers) that is originally registered to Tolman, Jampes P. on May 27, 1884. The last listed owner is Samson Ocean Systems, Inc. *Samson* has a design and word mark. The design is shown on *TESS*. Design search codes describe the design.

A similar query is: (`fd < 18900000) and (`rn > "0") and live [ld]

This results in seven marks. The author notices that there is one irrelevant registration with a filing date of 0000.

To find other registrations by the same owners, both owner names are used in the strategy:

(samson adj ocean adj systems)[on] or (tolman near1 jampes) [on]

This shows some records with Samson Cordage Works and American Cordage as the last owners. These names can also be added to the search strategy.

b. In *Trademarks BIB* the Show Index feature works well for this query. The author's theory is that the software is running on a standalone computer with no competing users. This is how it is used:

The date should be 1870 or later. A query may be limited to active registrations by using the Live-Dead Indicator and searching for "registered." The choices for Live-Dead Indicator are registered, pending, and dead.

The form search is easier for this query. Use the pull-down menus to identify the fields: Live-Dead Indicator and Registration Date. At the Live-Dead Indicator, one types the word "registered." To use Show Index, one selects the data entry area for Registration Date and selects the Show Index Icon. A display showing dates appears. A date, e.g., 1870 is entered and the first date that appears is 18840527. This date is selected and automatically entered in the Registration Date field.

| Live-Dead Indicator | Registered | AND |
| Registration Date | 18840527 | |

The Search Icon is selected and one item is found.

With command search, the Show Index feature is also helpful. A similar operation is done:

((registered)).LD. AND ((@RD = "18840527"))

The result shows that Samson is trademarked for the original registrant, Tolman, Jampes P. and the last listed owner, Samson Ocean Systems, Inc. One finds many other trademarks by the same owner. Phrase searching speeds this suggested strategy:

"samson ocean"[on] or tolman near1 jampes[on]

When one searches for Samson Ocean Systems marks, one realizes that Samson Cordage and American Cordage are related owners. When these two companies are added, one finds 31 trademarks.

Printing and exporting results in the default format is very helpful for most search results. One views the display style and customizes the fields. In this case, the author suggests Registration Date, Registration Number, Filing Date, Serial Number, Word Mark, Design Code, Owner Name, and Owner Address. These fields should provide the basic information that can be printed or exported as needed. Goods and Services and International Class are also key fields. Some exporting problems may result when multiple Owners, Goods and Services, and International Classes are contained in a single record.

c. In *Patents and Trademarks ASSIGN*, one tracks ownership changes by searching assignor or assignee and document type. The field tags are:

    Assignor   .AR.
    Assignee   .AE.
    Document Type   .DT.

In command search, this strategy should work:

(("samson ocean" or "samson cordage" or "american cordage" or "tolman near2 jampes").AR. or ("samson ocean" or "samson cordage" or "american cordage" or "tolman near2 jampes").AE.) and trademark.DT..

Eight records are found. Several have many marks transferred in a single record. Printing records is recommended since exporting the records presents problems when multiple marks are contained in a single record.

d. In *USAMark*, it is possible to print registration certificates and mark images. One retrieves the certificate by its Registration Number. These numbers may be exported from *Trademarks BIB*.

Query 3

XYZ Police Vests, a fictitious applicant, is filing a mark for bullet-proof vests and wants to avoid choosing a mark that might be confused with a current mark. XYZ wants to see the broad picture of marks that are currently used so that their company can choose a unique mark, save marking costs, and avoid future litigation. This search focuses on retrieving all marks for the Goods involved. The *Trademark Acceptable Identification Goods and Services Manual* lists "bullet-proof vests" and "bullet-proof vests and clothing." These are from International Class 009 that includes "life-saving and teaching apparatus and instruments."[34] It may be helpful to find a known mark, e.g., Kevlar, to get more ideas for identifying goods. This example shows the creativity that is needed to identify goods: police vests, body armour, body armor, protective clothing, bullet-proof vests, ballistic vests, bullet resistant vests, and flack jackets. One may choose to search terms in the goods and services field or search the terms without naming a field. More records are retrieved if a field is not named. To do a thorough search, it may be necessary to view some irrelevant marks.

a. Table 8 shows strategies used in the *Trademarks Electronic Search System (TESS)*.
b. In *Trademarks BIB*, similar searches are done in command search. One substitutes the *TESS* field tag format [gs] for BIB format of .gs. It is also possible to search the whole record when not naming a field. Compared with *TESS*, the results are slightly smaller because less time is covered by BIB. If one wants to sort or export date, BIB is better than *TESS*.

Query 4

The eBay Trading Post Company, which runs an online auction and flea market service, is concerned that other companies may be attempting to register marks that might be confusingly similar to its own. The Services are related to online bidding. Searches may be done on *TESS* and *Trademarks BIB*. *TESS* may be the preferred system for this query since one is looking for up-to-date information. The *Official Gazette–Trademarks* may also be consulted for marks published for opposition.

Trademark Acceptable Identification of Goods and Services Manual lists "On-line trading services in which seller posts products to be auctioned and bidding is done via the Internet" in Service Class 035. This

TABLE 8. Search Strategies Used in *TESS* for Bullet Proof Vests

Search Strategy	Results
((bullet adj proof or bulletproof) and vest$1)[gs]	116 records
((bullet or bullets or bulletproof or ballistic$ or police or flack) and (clothing or vest$1 or jacket$1) or body adj armour or body adj armor)[gs]	492 records
((bullet or bullets or bulletproof or ballistic$ or police or flack) and (clothing or vest$1 or jacket$1) or body adj armour or body adj armor)	660 records
((bullet or bullets or bulletproof or ballistic$ or police or flack) and (clothing or vest$1 or jacket$1) or body adj armour or body adj armor) and "009"[ic]	316 records

international class covers the broad category of "Advertising; business management; business administration; office functions." This class may not be the only one that one wants to search since eBay may consider creating clothing and other products that shows its mark.

Searching word marks usually involves the basic index or combined fields. The field tags are [BI] or [COMB]. Field tags may be entered in upper or lower case. The basic index includes the mark non-punctuated, mark punctuated, and pseudo mark. The translation index is searched separately as [ti]. Table 9 defines these fields.

New User Form Search (Basic) uses the field Combined [COMB]. Free Form Search (Advanced Search) uses the Basic Index [BI]. *TESS Help* indicates that the Basic Index "contains the word mark and pseudo mark information indexed for optimal searching efficiency." Translation is not included in the Basic Index. Truncation symbols vary slightly from the $ depending on the field. Basic Index allows different truncation than does Translation. Relevant help menus explain truncation and pattern matching. Strategies may be simple or complex and may involve truncation and pattern matching particularly in the word mark related fields.

a. Table 10 shows examples from searches of *TESS* in Advanced Search.
b. Although similar searches can be done on *Trademarks BIB*, one realizes that the data is two to three months old. The truncation symbols vary so the author recommends review of the help screens. Many marks may be pending for some time so *Trademarks BIB* is appropriate for pending marks. If one is looking only for marks that have been published for opposition, *TESS* is the preferred system.
c. The print or online *Official Gazette–Trademarks* may also be used to review marks that are recently published for opposition.

TABLE 9. Definitions for Word Mark Fields in *TESS* for Query 4 *(Exact Quotes or Adapted from* TESS Help: *Search Fields*[1]*)*

Field Name	Definition	Unique Field Tag	Basic Index Field Tag
Mark Punctuated	Word Mark, "including any punctuation characters. This is the word mark field displayed in the text output."	[MP]	[BI] or [COMB]
Mark Non-Punctuated	Text for the Word Mark with punctuation characters removed and replaced by a space.	[MN]	[BI]
Pseudo Mark	Word Mark that contains an "alternative or intentionally corrupted spelling for a normal English word." This field "often contains spellings that are very similar or phonetically equivalent to the Word Mark." It is added by USPTO staff and is not displayed in *TESS record*.	[PM]	[BI]
Translation Statement	"English equivalents to foreign words or characters used in a trademark"	[TL]	[COMB]
Translation Index	"English equivalents to foreign words or characters used in a trademark"	[TI]	Not applicable

[1] United States Patent and Trademark Office. *Trademark Electronic Search System Help: Search Fields. 3/6/03.* Accessed *on* 6 March 2003. <http://tess2.uspto.gov/bin/gate.exe?f=help&state=ce08pg.1.1#SearFiel>

Query 5

An independent inventor is looking for companies that might license his patented invention for integrated television and Internet system. The person does not have the financial resources or technical expertise to bring this product to market. Someone interested in this query may typically search patent databases to identify licensing opportunities. Searching for trademark owners who produce or sell similar goods may be another source of information. It is possible to identify trademark owners from *TESS* or *Trademarks BIB*.

Owners may be searched as patent assignees in the Patent Full-Text and Full-Page Image Databases[35] on the USPTO Web site. These include assignees from 1976 to date for Issued Patents and assignees from 2001 to date for Patent Applications. In the *Cassis2* Patents BIB database on DVD, assignees are available from 1969 to date. Searching by

TABLE 10. Sample Searches on eBay

Simple search	(ebay or e adj bay or ebae)[bi,ti]
Left-hand and right-hand truncation	*bay*[bi] or *bae*[bi]
Internal truncation with any letter	*b?y*[bi]
Internal truncation with any letter and Goods or Services	*b?y*[bi] and (online or on adj line or Internet or web)[gs] same (trade or trades or trading or bid or bids or bidding or auction$)[gs]
Truncation with any letter, Goods or Services, and recently filed applications	*b?y*[bi] and (online or on adj line or Internet or web)[gs] same (trade or trades or trading or bid or bids or bidding or auction$)[gs] and `FD > "20020100"
Truncation with any letter, Goods or Services, and application recently published for opposition	*b?y*[bi] and (online or on adj line or Internet or web)[gs] same (trade or trades or trading or bid or bids or bidding or auction$)[gs] and `PO > "20021000"
Vowel pattern patching, Goods or Services, and International Class 035	(*b{v}y*[bi] and (online or on adj line or Internet or web)[gs] same (trade or trades or trading or bid or bids or bidding)[gs] and 035[ic]

Current Classification (Class/Subclass) may also be used to identify other interesting patents.

Both *TESS* and *Trademarks BIB* are helpful for this search. When one searches, the most useful field may be Goods and Services:

a. In *TESS*, this strategy may retrieve some relevant owners:

television[gs] near7 integrat$[gs] and (web or webs or Internet)[gs]

This results in 67 records. Relevant trademark owners may be identified for possible licensing opportunities from these records.

b. In *Trademarks BIB*, a similar search may be done:

television.gs. near7 integrat$.gs. and (web or webs or Internet).gs.

The sorting, printing, and exporting capabilities of *Trademarks BIB* have some advantages for this search.

REVIEW OF QUERIES

Given the previous queries, the author realizes that both systems can provide good results for most queries. It appears that no one system can answer all queries. It is recommended that the searcher examine the strengths and weaknesses of each system and determine which may be the best for each query. It may be necessary to search both systems in some cases.

CONCLUSION

Knowledge of trademark searching terminology and process better prepares clients to use trademark databases from the United States Patent and Trademark Office. Understanding the strengths and weaknesses of Web and *Cassis2* databases makes it possible to use the most appropriate one to answer queries. Acquiring skills in using specific fields make clients and librarians better prepared to explore the databases.

The author encourages librarians and trademark clients to use a variety of trademark databases to answer new queries on competitive intelligence, job seeking, infringement cases, licensing opportunities, trivia, or history. The author hopes that these examples encourage clients and librarians to find creative uses of trademark data.

NOTES

1. *Cassis2* has its origins in *Cassis (Classification and Search Support Information System)* CD-ROM series which was first provided to all Patent and Trademark Depository Libraries in 1987 using Dataware CD Answer software. U.S. Patent and Trademark Office. *1999 Products and Services Catalog. Washington, D.C.: U.S. Government Printing Office, 1999. Page 17.*

2. In mid-1999, the *Cassis* software was replaced by Dataware II (aka CD Publisher) which runs on Windows 95/98, 2000 or the NT platform and is called *Cassis2*. "*Cassis2* consists of six basic DVD-ROM databases containing various types of patent and trademark data, and a series of DVD-ROMs containing facsimiles of patent and trademark documents. All are available for purchase from the USPTO." After several company changes, the *Cassis2* software appears to be produced by Open Text. This is currently displayed on *Cassis2 DVDs*. 4/12/2002. Accessed on 5 March 2003. <http://www1.uspto.gov/web/offices/ac/ido/oeip/catalog/products/cassis.htm>.

3. United States Patent and Trademark Office. *Basic Facts about Trademarks.* 10/16/01. Accessed on 11 August 2002. < http://www.uspto.gov/web/offices/tac/doc/basic/>.

4. United States Patent and Trademark Office. *Frequently Asked Questions about Trademarks.* 6/24/02. Accessed on 11 August 2002. <http://www.uspto.gov/web/offices/tac/tmfaq.htm>.

5. Mary Hannon, ed. *Trademark Manual of Examining Procedure (TMEP).* 3rd ed. Revision 1. Washington, D.C.: U.S. Patent and Trademark Office, June 2002. 6/24/02. Accessed on 25 July 2002. <http://www.uspto.gov/web/offices/tac/tmep/>.

6. Ibid. "What is a trademark or service mark?" *Basic Facts about Trademarks.* 10/16/2001. Accessed on 6 March 2003. <http://www.uspto.gov/web/offices/tac/doc/basic/trade_defin.htm>.

7. Ibid. "What Must the Application Include? Basis for Filing: What is use in commerce?" *Basic Facts About Trademarks.* 12/09/2002. Accessed on 23 February 2003. <http://www.uspto.gov/web/offices/tac/doc/basic/appcontent.htm>.

8. Ibid. "What Constitutes Interstate Commerce?" *Frequently Asked Questions about Trademarks.* 2/12/2003. Accessed on 6 March 2003. <http://www.uspto.gov/web/offices/tac/tmfaq.htm#Basic010>.

9. Michael White. *State Trademarks.* 7/7/2002. Accessed on 27 February 2003. <http://statetm.tripod.com/>.

10. Kevin R. Harwell. "Resources for Searching Common Law Trademarks." *Reference & User Services Quarterly* 39, no.4 (Summer 2000): 336-341.

11. "What Must the Application Include? Basis for Filing," *Basic Facts about Trademarks.* 6/25/02. *Accessed on 23 September 2002.* <http://www.uspto.gov/web/offices/tac/doc/basic/appcontent.htm#basis>.

12. Ibid. "Definitions: What is a trademark? What is a service mark? What is a certification mark? What is a collective mark?" *Frequently Asked Questions About Trademarks.* 2/27/03. Accessed on 27 February 2003. <http://www.uspto.gov/web/offices/tac/tmfaq.htm#DefineTrademark>.

13. United States Patent and Trademark Office. *Trademark Acceptable Identification of Goods and Services Manual.* 11/06/2002. Accessed on 6 March 2003. <http://www.uspto.gov/web/offices/tac/doc/gsmanual/>.

14. Countries that sign the Nice Agreement agree to adopt a common classification system for goods and services. "The Classification consists of: (i) a list of classes, together with, as the case may be, explanatory notes; (ii) an alphabetical list of goods and services (hereinafter designated as "the alphabetical list") with an indication of the class into which each of the goods or services falls." *The Nice Agreement Concerning the International Classification of Goods and Services for the Purposes of Registration of Marks of June 15, 1957, as revised at Stockholm on July 14, 1967, and at Geneva on May 13, 1977, and amended on September 28, 1979.* Accessed on 23 February 2003. <http://www.wipo.int/clea/docs/en/wo/wo019en.htm>.

15. Phyllis Karrh and Robin Kelley, "Trademarks: More Than Meets the Eye." *Indiana Libraries* 19, no.1 (2000): 28-30.

16. Ibid. "Can the Office Refuse to Register a Mark?" *Frequently Asked Questions about Trademarks."* 2/12/2003. Accessed on 27 February 2003. <http://www.uspto.gov/web/offices/tac/tmfaq.htm#Application011>.

17. U.S. Patent and Trademark Office. *Official Gazette of the United States Patent and Trademark Office (Trademarks).* [Washington, D.C.] : U.S. Dept. of Commerce, Patent and Trademark Office: [Superintendent of Documents, U.S. Government Printing Office, distributor], 1975-date.

18. United States Patent and Trademark Office. *Trademark Official Gazette.* 10/01/02. Accessed on 04 October 2002. <http://www.uspto.gov/web/trademarks/tmog/>.

19. Ibid. *Trademark Manual of Examining Procedure (TMEP).* 803.01.

20. Ibid.

21. United States Patent and Trademark Office. *Trademark Electronic Search System Help: Introduction. 3/4/2003. Accessed on 3/4/2003.* <http://tess2.uspto.gov/bin/gate.exe?f=help&state=68elhn.1.1#Intro>.

22. On March 4, the last complete filing date for paper applications was 01/01/2003 which was loaded on 01/23/2003 and the last complete filing date for e-TEAS, i.e., electronic applications, was 1/20/2003 which was loaded on 2/19/2003. United States Patent and Trademark Office. *Trademark Electronic Search System News.* 2/19/2003. Accessed on 4 March 2003. <http://tess2.uspto.gov/webaka/html/news.htm>.

23. United States Patent and Trademark Office. "Limitations Regarding The TARR Database." *Trademark Applications & Registration Retrieval (TARR).* 2/19/2003. Accessed on 6 March 2003. <http://tarr.uspto.gov/disclaimer.html>.

24. United States Patent and Trademark Office. *Products and Services On-Line Catalog: Trademark Products.* 2/26/2003. Accessed on 27 February 2003. <http://www.uspto.gov/web/offices/ac/ido/oeip/catalog/products/tmprod-1.htm>.

25. United States Patent and Trademark Office. *USPTO Design Search Code Manual.* 11/26/2002. Accessed on 11 August 2002. <http://tess2.uspto.gov/>.

26. Ibid. "International Schedule of Classes of Goods and Services." *Basic Facts About Trademarks.* 2/27/2003. Accessed on 27 February 2003. <http://www.uspto.gov/web/offices/tac/doc/basic/international.htm>.

27. Ibid. *Trademark Acceptable Identification of Goods and Services Manual.*

28. Patents. "Assignment, Recording and Rights of Assignee: Definitions." *Code of Federal Regulations.* 37 C.F.R. §3.1. (Washington, D.C.: U.S. Government Printing Office, 2001), 223.

29. Ibid. *Trademark Acceptable Identification Goods and Services Manual.*

30. Ibid. "International Schedule of Classes of Goods and Services." *Basic Facts about Trademarks.*

31. United States Patent and Trademark Office. "Frequently Asked Questions: What does the search result Word Limit Overview mean?" *Trademark Electronic Search System Help.* 3/6/2003. Accessed on 6 March 2003. <http://tess2.uspto.gov/bin/gate.exe?f=help&state=1p09pk.1.1#FAQ_WLO>.

32. Ibid. *Trademark Electronic Search System Help: Search Fields.* 3/5/2003. Accessed on 5 March 2003. <http://tess2.uspto.gov/bin/gate.exe?f=help&state=f75qr6.1.1#SearFiel>.

33. U.S. Trademark Registration 0011210. *Samson.* Tolman, Jampes P. Dated May 27, 1884. 5th Renewal, September 1, 1994. Expires September 1, 2004.

34. Ibid. "Goods: 9 . . . 'life-saving and teaching apparatus and instruments.' International Schedule of Classes of Goods and Services." *Basic Facts About Trademarks.* 2/27/2003. Accessed on 27 February 2003. <http://www.uspto.gov/web/offices/tac/doc/basic/international.htm>.

35. United States Patent and Trademark Office. *Patents Full-Text and Full-Image Databases: Issued Patents and Patent Applications.* Washington, D.C.: USPTO, 1790-date. 9/25/02. Accessed on 4 October 2002. <http://www.uspto.gov/patft/index.html>.

State Trademark and Company Name Web Sites

James C. Miller

SUMMARY. This paper briefly discusses the importance of searching state trademarks and reviews the library and legal literature on issues involved in such a search. It summarizes the development of search services, from Dialog's release of *Trademarkscan® State* in January 1987 to the present proliferation of individual states' free company name and trademark databases on the Internet. It describes commonly available search features and unusual special features, and compares their availability in different states' web sites. A spreadsheet gives details and web addresses for the fifty states, Puerto Rico, and the District of Columbia. *[Article copies available for a fee from The Haworth Document Delivery Service: 1-800-HAWORTH. E-mail address: <docdelivery@haworthpress.com> Website: <http://www.HaworthPress.com> © 2001 by The Haworth Press, Inc. All rights reserved.]*

KEYWORDS. State trademarks, state trademark databases, state company name databases

THE IMPORTANCE OF STATE TRADEMARK SEARCHING

The business and legal literature has featured a number of articles that strongly advise searching state and common law trademarks and

---

James C. Miller is Senior Reference Librarian and College Park PTDL Representative, Engineering & Physical Sciences Library, University of Maryland, College Park, MD 20742-7011 (E-mail: jm69@umail.umd.edu).

[Haworth co-indexing entry note]: "State Trademark and Company Name Web Sites." Miller, James C. Co-published simultaneously in *Science & Technology Libraries* (The Haworth Information Press, an imprint of The Haworth Press, Inc.) Vol. 22, No. 1/2, 2001, pp. 161-173; and: *Patent and Trademark Information: Uses and Perspectives* (ed: Virginia Baldwin) The Haworth Information Press, an imprint of The Haworth Press, Inc., 2001, pp. 161-173. Single or multiple copies of this article are available for a fee from The Haworth Document Delivery Service [1-800-HAWORTH, 9:00 a.m. - 5:00 p.m. (EST). E-mail address: docdelivery@haworthpress.com].

http://www.haworthpress.com/store/product.asp?sku=J122
© 2001 by The Haworth Press, Inc. All rights reserved.

names. Ojala (1996) notes that some business researchers think only legal professionals should do trademark searching. She sees no problem with searching before spending money for legal assistance, as long as you realize the limits of this search. Gathering intelligence on other companies, looking for infringement of your existing trademarks, and judging whether a prospective mark poses clear problems, are all valid reasons for doing a prior search. Thompson (2001) cites "Hilfiger's choice not to perform a full search," resulting in an unfavorable decision in *International Star Class Yacht Racing Association v. Tommy Hilfiger*, 80 F.3rd 749, 753 (2nd Cir. 1996). Gindin (1998) sees the proliferation of databases and web sites making "increasing responsibility as well as opportunity for the trademark searcher."

Duboff (1994) emphasizes that States do not check pending state applications in any federal trademark databases, so state trademark and company name databases are critical places to watch in protecting your trademark. Rector (1999) discusses reasons for getting a state trademark, such as early protection while awaiting Federal registration, the benefits of individual state laws to prosecute infringement, and having more than simply common law protection even if a Federal trademark is denied on various grounds. Sanford (1998) compares applying for a trademark without prior searching to buying or leasing property without a proper title search. The Internet is at least a cheaper second choice if you cannot afford a professional search.

Holliday (1993) discusses trademark issues faced by bankers and other financial providers, and devotes most of a page to state registration. She notes that large industrial states such as New York tend to have strong anti-dilution laws, demanding extra care in avoiding any similarity to parts of other marks. Elizabeth Yahiel (1995), Corporate and Trademark Attorney for the Texas Secretary of State, Corporations Section, describes Texas trademark law in detail, and lists ninety-six references to specific cases and to the Texas code.

THE DEVELOPMENT OF STATE TRADEMARK DATABASES

Although professional search services have included state, common law, and even international trademark searching for a long time, most librarians probably first heard of state trademark searching as part of the *Trademarkscan®* database. Dialog® first announced *Trademarkscan® State* in December 1986 and made it available in January of 1987 (*Chronolog*). From the beginning, it included fifty states and Puerto Rico, but its Update Table in *Chronolog* January 1987 noted only trade names for Arizona and service marks for New York. *Database Magazine* (1994) reported Thomson & Thomson's "second generation" of *Trademarkscan® State* database, including the addition of 300,000 records. Berinstein (1997) listed *Lexis-Nexis* files with images, including the state trademarks (*STTM*) database provided by CCH Trademark Research Corporation. More recently, Hurst (2000) and Quint (2000) reviewed *Trademark.com*, a fee-based database from Information Holdings, Inc. subsidiary *CorporateIntelligence.com*. It started in 2000 with state, federal, and 800,000 common law marks.

TYPE OF WEB ACCESS OFFERED BY THE 50 STATES AND PUERTO RICO

Internet access to state trademarks has developed quite rapidly. Harwell (2000) found about fifteen states offering free trademark search. As of September 2002, twenty-six states have web databases of registered companies, and twenty-one more offer both company and trademark searching for free. Five states do not yet offer free web searching of either trademarks or company names but will do searches by request, some for a small fee. With a few exceptions, state trademarks and corporation names are registered with the Secretary of State. Michael White (2002) of the U.S. Patent & Trademark Office's Patent & Trademark Depository Library Program has compiled links at http://statetm.tripod.com. These databases change often. To check broken links, first try the state's main homepage, which is usually in the format http://www.state.xx.us, where *xx* is the state's abbreviation. If the Secretary of State does not have a database, check the state's Corporations or Business department or division. Figure 1 shows which states have trademark or company name databases, both kinds of databases, or no free databases. The Appendix gives details of databases and features for each state and includes web addresses as of September 2002.

States with Both Trademark and Company Name Search

Alaska, Arizona, Arkansas, Colorado, Florida, Georgia, Hawaii, Kentucky, Louisiana, Maine, Maryland, Minnesota, Montana, New Jersey, North Dakota, Ohio, Tennessee, Utah, and Vermont all offer free company name and trademark searches. Virginia offers free searching of both on its State Corporation Commission web site, but you have to fill out an online registration form to search companies. Michigan offers both, but trademarks are in a PDF (Portable Document Format) file that must be searched using the "Find in Page" feature of your browser.

FIGURE 1. Databases Available

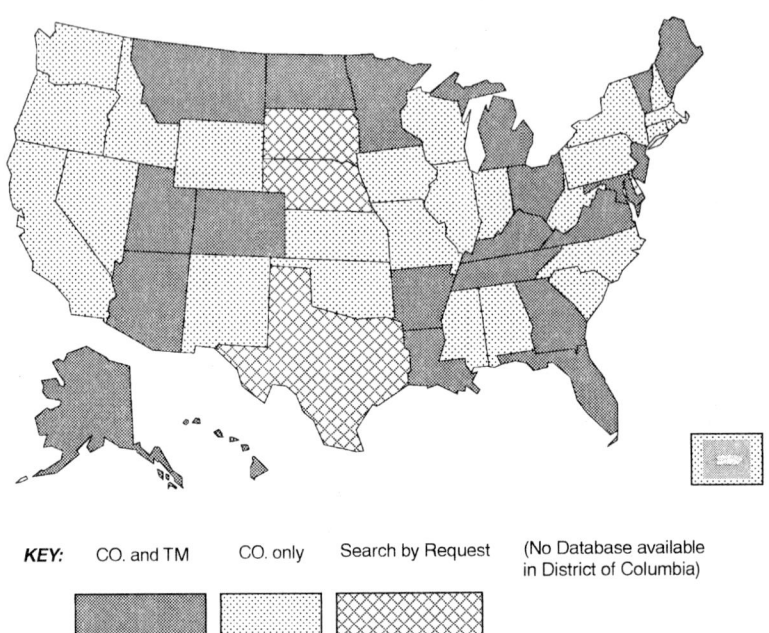

KEY: CO. and TM   CO. only   Search by Request   (No Database available in District of Columbia)

States with Only Company Name Search

The twenty-six states offering only company searches for free are Alabama, California, Connecticut, Idaho, Illinois, Indiana, Iowa, Kansas, Massachusetts, Mississippi, Missouri, Nevada, New Hampshire (companies with web sites only), New Mexico, New York, North Carolina, Oklahoma, Oregon, Pennsylvania, Puerto Rico, Rhode Island, South Carolina, Washington, West Virginia, Wisconsin, and Wyoming. In 2001 Puerto Rico made available monthly PDF listings of trademarks with color images (*Avisos Marcas*) and corporations (*Nombres Comerciales*). For a $5 fee, Oklahoma has a "Business Records" search that includes trade names. West Virginia also will do trademark searches for a small fee.

States Without Free Web Databases for Either Companies or Trademarks

Texas offers both trademark and company name searches for a nominal $1 fee. Delaware and Nebraska also offer both for a small fee. South

Dakota will do searches on request, charging only for copies. The District of Columbia has no company or trademark databases. New Hampshire will search its UCC database for a fee, but seems to have no trademark information.

TYPES OF SEARCH ENGINES AND SPECIAL FEATURES

Figure 2 gives the number of states that have common search functions, and Figure 3 shows how many states offer more unusual features.

Eighteen states with both browse and keyword search are Alabama, Hawaii, Illinois, Kansas, Kentucky, Massachusetts, Michigan, Missouri, New York, North Carolina, North Dakota, Oklahoma, Oregon, Rhode Island, South Carolina, Vermont, Virginia, and Washington. Thirteen allow browse only: Alaska, Florida, Iowa, Louisiana, Minnesota, Mississippi, Montana, New Jersey, New Mexico, Pennsylvania, West Virginia, Wisconsin, and Wyoming. Sixteen allow keyword only: Arizona, Arkansas, California, Colorado, Connecticut, Georgia, Idaho, Indiana, Maine, Maryland, Nevada, New Hampshire, Ohio, Puerto Rico (companies only), Tennessee, and Utah.

Thirteen states provide phrase searching: Hawaii, Idaho, Kansas, Massachusetts, Nevada, New York, Oregon, Rhode Island, South Carolina, Tennessee, Utah, Virginia, and Washington. Thirty-nine states allow some form of truncation, all but: Alabama, California, Connecticut, Delaware, DC, Iowa, Minnesota, Montana, Nebraska, New Mexico, Pennsylvania, South Dakota, and Texas. Seven of these thirty-nine require the % truncation symbol, or the choice of "stem" or "partial" words on their search menus: Arizona, Maine, New Jersey, New York, North Dakota, Ohio, and Virginia.

FIGURE 2. Search Features *(No. of States)*

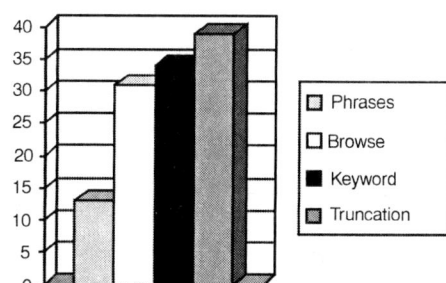

FIGURE 3. Unusual Features *(No. of States)*

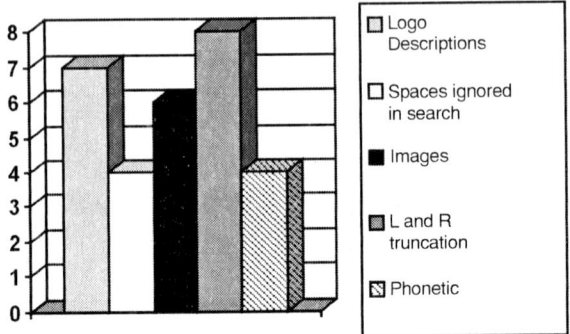

UNUSUAL FEATURES, IMAGES AND DESCRIPTIONS

Arizona, Colorado, Massachusetts, and Oregon have search engines that feature phonetic or "sound-alike" searching. Georgia, Maine, Maryland, New Hampshire, New Jersey, Oklahoma, Rhode Island, and Virginia allow left and right truncation. Four states' search engines ignore spaces (*redro* retrieves *red rose, red roof*, for example): Florida, Iowa, Maine, and New Jersey.

Colorado, Florida, North Carolina, Ohio, Puerto Rico, Tennessee, and Virginia offer at least some images. Puerto Rico's monthly "Avisos Marcas" has many color images and Tennessee and Virginia databases have a few in color. Colorado has file histories in .PDF format, and North Carolina has page images of company reports online. Arizona, Arkansas, Georgia, Kentucky, Minnesota, North Dakota, and Utah have descriptions of trademark images, but not all allow searching of these descriptions.

CONCLUSION

Most states provide relatively simple interfaces. The thirty-nine that allow or default to truncation recognize that a prospective name is a risky choice if it sounds similar to an existing name or part of that name. Many sites plainly warn the searcher to look out for names with similar sounds. Most sites' search features tend to be limited. However, during the year's duration of this study, many sites have enhanced their search engines and coverage. Because paper files are very expensive, we can expect many more sites to add images to their databases. Existing trademarks databases should help spread awareness of state marks as a cheaper alternative to federal registration, and encourage more states to offer their own trademark sites.

REFERENCES

Berinstein, Paula. "Lexis-Nexis: images that contribute to the bottom line." *Online* 21:3 (May/June 1997): 79-80.

Chronolog. (Dialog® File 410, searched 5/2/02). December 1986, January 1987.

Duboff, Leonard D. "What's in a name: the interplay between the federal and trademark registries and state business registries." *DePaul Business Law Journal.* 6 (Winter 1994): 15 (see end of section B "Protection under Federal Trademark Law").

Gindin, Susan E. "Researching trademarks." http://www.info-law.com/tmsearch.html on her law practice web site at http://www.info-law.com/, accessed 5/21/02. Update (1998) of article in *Legal Information Alert*, October 1995 ("Conclusion," p. 13).

Harwell, Kevin R. "Resources for searching common law trademarks." *Reference and User Services Quarterly* 39:4 (Summer 2000): 336-41. Also in *WilsonSelect Full Text* (searched 2/02).

Holliday, Karen Kahler. "How to avoid TM troubles." *Bank Marketing* 25: 5 (May 1993): 18-21.

Hurst, Jill Ann. "The new kid on the trademark block: Trademark.com." *Econtent* 23:5 (Oct/Nov 2000): 51-54. In *WilsonSelect Full Text*.

Ojala, Marydee. "Trademarks for the business searcher." *Online* 20: 2 (Mar/Apr 1996):52.

Quint, Barbara. "Trademark.com: a new alternative for trademark searchers." *Information Today* 17:6 (June 2000): 30-31. Also at www.infotoday.com/newsbreaks/nb000508-1.htm (searched 5/25/02).

Rector, Susan D. "State trademark registration: to file or not to file?" *Intellectual Property Strategist.* 5: 11 (August 1999): 8.

Sanford, Gordon U. "Intellectual property roadmap: the business lawyer's role in the realm of intellectual property." *Mississippi College Law Review.* 19 (Fall 1998): 177. (see section d.ii, "The registration process" quoting Field, Thomas G. in "Avoiding intellectual property problems" on Franklin Pierce Law Center web site http://www.PIERCELAW.EDU/tfield/aVoid.htm modified May 12, 2001).

Thompson, Laura. "Part three: The registration process: Due diligence: trademark searches and opinion letters." *Journal of Contemporary Legal Issues.* 12 (2001):105. (see note 6).

White, Michael. *State Trademarks*. Revised July 7, 2002. http://statetm.tripod.com/.

Yahiel, Elizabeth B. "Process and review of state trademark applications with the Office of the Secretary of State of Texas." *Texas Intellectual Property Journal* 67:3 (Winter 1995). In *Lexis-Nexis Academic Universe* 11419 words (20 p.).

APPENDIX. State Trademark Databases and Features

STATE	INFORMATION	SEARCH FEATURES	Key	BR	PHR	TRUNC	OTHER	URL	
Alabama	corporations, officers	browse index, keyword	y	y	n	n		***Search corporations:*** http://www.sos.state.al.us/business/corporations.cfm ***Trademark forms and information:*** http://www.sos.state.al.us/business/land.htm	co
Alaska	corporations, officers, trademarks	browse index, search by class	n	y	n	auto	class (US?) search	***Search trademarks:*** http://www.dced.state.ak.us/bsc/TrdStart.cfm ***Search corporations:*** http://www.dced.state.ak.us/bsc/CorporationSearch.cfm	both
Arizona	corporations, officers, trademarks (including description of mark) in one database	keyword default=AND; exact, partial, soundex, metaphone words	y	n	n	menu	soundex metaphone. Descrip.	***Search corporations, agents, trademarks:*** http://www.sosaz.com/scripts/TNT_Search_engine.dll	both
Arkansas	corporations, cooperatives, officers, trademarks description of TM images	keyword, returns marks in alphabetic order	y	n	n	auto	descrp.	***SearchTrademarks:*** http://www.sosweb.state.ar.us/corps/trademk Main search menu including ***incorporations, cooperatives, non-profits:*** http://www.sosweb.state.ar.us/corps/	both
California	they will search; 80,000 marks in trademarks database. Corporations can be searched	trademarks by request. corporations by keyword, no phrases (space means AND)	y	n	n	n	80,000 records in Trademark database	***Trademark information:*** http://www.ss.ca.gov/business/ts/ts.htm ***Search corporations:*** http://kepler.ss.ca.gov/list.html	co
Colorado	mixed TR, business names in one database, file histories available in .pdf, some images	browse index, phonetic; search "design only" for examples of images	y	n	n	auto	phonetic, images	***Companies and trademarks:*** Choose "***Searchable Database of business records ...***" at: http://www.sos.state.co.us/pubs/business/main.htm More information on ***trademarks:*** http://www.sos.state.co.us/pubs/business/faq_bus.html#_Toc17680499	both
Connecticut	company names; they will search trademarks ($25, pdf form on web) Tel: 860-509-6003 crd@po.state.ct.us	keyword, limited info on web, menu buttons	y	n	n	n	non-Windows interface-menu buttons	***Company name search:*** http://www.concord.state.ct.us/servlet/TerminalServlet?request=send&transaction=SSRM ***TM info at:*** http://www.sots.state.ct.us/CommercialRecording/CRDform.html#Trade and Service Mark	co
Delaware	they will search trademarks, company names. Info at tel. (302) 739-3073	no web database 11/03						***Info at:*** http://www.state.de.us/corp/regguide.shtml *and* http://www.state.de.us/corp/nrdetail.shtml	none

								Corporation registration info at: http://dcra.dc.gov/information/build_pla/business_services/coporations_division.shtm	none
District of Columbia	No database available 7/01	no web database 11/03							
Florida	trademarks, company names. Some images	browse index	n	y	n	y		Search Companies, Trademarks: http://ccfcorp.dos.state.fl.us/corpweb/inquiry/cormenu.html	both
Georgia	trademarks, mark descriptions, company names	keyword, automatic left and right truncation	y	n	n	y	ignores spaces. Images	Search Trademarks: http://www.SOS.State.Ga.US/corporations/marksearch.htm Search Companies: http://www.SOS.State.Ga.US/corporations/corpsearch.htm	both
Hawaii	trademarks, company names in one file, "certificate of good standing" in separate file	browse index, keyword or phrase, but no Boolean, automatic truncation	y	y	y	auto	descrip.	Search Trademarks and Companies: http://www.ehawaiigov.org/DCCA/biz-name/html/ "Certificate of Good Standing": http://www.ehawaiigov.org/DCCA/cog/html/	both
Idaho	company names with document images	keyword, phrase, automatic truncation	y	n	y	auto	some phrases loaded as keywords	Search companies: http://www.accessidaho.org/apps/sos/corp/search.html Trademark Info: http://www.idsos.state.id.us/tmarks/tmindex.htm	co
Illinois	company names. Weekly updates	browse, keyword, no phrases, automatic truncation	y	y	n	auto		Search Corporations: http://www.sos.state.il.us/cgi-bin/business_services/corpsrch.s TM Information and forms: http://www.cyberdriveillinois.com/departments/business_services/trademrk.html	co
Indiana	company names	keyword, automatic truncation; no phrases (space = AND)	y	n	n	auto		Search Companies: https://www.ai.org/sos/bus_service/online_corps/default.asp Name Availability: https://www.ai.org/sos/bus_service/online_corps/nameavail.asp TM info: http://www.IN.gov/sos/business/trademarks.html	co
Iowa	company names	browse index, ignores spaces	n	y	n	n	ignores spaces	Search Companies: http://www.sos.state.ia.us/corp/corp_search.asp Download forms, incuding TM/SM application: http://www.sos.state.ia.us/business/form.html	co
Kansas	company names	browse, keyword, agent names, automatic truncation	y	y	y	auto		Search Companies: http://www.accesskansas.org/corporations/ TM and SM forms: http://www.kssos.org/forms/forms_results.asp?division=BS#Trade-marks/Service-marks	co
Kentucky	company names	browse index, automatic truncation	y	y	y	auto		Search Companies: http://www.sos.state.ky.us/corporate2/entityname.asp TM info at: http://www.kysos.com/ADMIN/LEGAL/trademarks.asp	co
Louisiana	trademarks, company names. Active in boldface	browse index, automatic truncation some images	n	y	n	auto	active in boldface	Search Companies & Trademarks: http://www.sec.state.la.us/crping.htm	both

APPENDIX (continued)

STATE	INFORMATION	SEARCH FEATURES	Key	BR	PHR	TRUNC	OTHER	URL	
Maine	trademarks, companies, names (choose from 29 categories or search all)	keyword, % is truncation, L-R Truncation ignores spaces & stop words. Can search mark desciptors	y	n	n	y	menu of 29 types of entities, L & R trunc. Ignores sp. Order copies $3-$8	**Search Companies, marks (choose from menu):** http://www.informe.org/icrs/ICRS?MainPage=x	both
Maryland	trademarks and companies in separate databases; company search is browse only	keyword, automatic left & right truncation or % to truncate company name, seems to search 1 word at a time	y	n	n	auto	auto L & R Trunc.	**Search Trademarks:** http://www.sos.state.md.us/sos/admin2/html/tmsearch.html **Search Business Entities:** http://sdatcert3.resiusa.org/ucc-charter/CharterSearch_f.asp	both
Massachusetts	company and individual name search	browse index, automatic truncation	y	y	y	menu, auto in full text		**Search Companies:** http://corp.sec.state.ma.us/corp/corpsearch/corpsearchinput.asp **TM info (no database):** http://corp.sec.state.ma.us/portal/Trademarks/TMMain.htm	co
Michigan	company and individual names; .pdf trademark listing (use "find on page")	browse index, automatic truncation, keyword	y	y	n	auto		**Search Companies:** http://www.cis.state.mi.us/bcs_corp/sr_corp.asp **Trademark .pdf listing:** http://www.cis.state.mi.us/bcsc/forms/corp/mark/markcom.pdf **Trademark Information:** http://www.cis.state.mi.us/dms/results.asp?docowner=BCSC&doccat=Mark&Search=Search	both
Minnesota	trademarks, including descriptions of logos; company names	browse index	n	y	n	n	more info. For $3 fee	**Search Companies, Trademarks (choose "search availability of a ... name"):** http://da.sos.state.mn.us/minnesota/home/dahome.asp **More business information:** http://www.sos.state.mn.us/business/index.html	both
Mississippi	company names, .pdf images of some documents	browse index, automatic truncation	n	y	n	auto on browse		**Search Companies:** http://www.sos.state.ms.us/busserv/corpsnap/index.html (No Trademark information)	co
Missouri	company names, agents	browse, keyword, automatic truncation	y	y	n	auto		**Search Companies and reserved names:** http://www.sos.mo.gov/BusinessEntity/soskb/csearch.asp **Trademark Information and forms:** http://www.sos.mo.gov/business/commissions/trademark.asp	co
Montana	trademarks, company names	browse index, no truncation	n	y	n	n	trunc only at beginning of name	**Search Companies and Trademarks:** http://app.discoveringmontana.com/bes/	both
Nebraska	company names	company keyword; request tradename search						**Search Companies:** https://www.nol.gov/sos/corp/corpsearch.cgi?nav=search **Trademark information:** http://assist.neded.org/tmark.html **Request trade name search:** http://assist.neded.org/tradname.html **Trademark, name, and company forms:** http://www.sos.state.ne.us/corps/	both - pay

State	Description	Search features			auto	finds ! + $ ... etc.	Search info	co/both
Nevada	company names, agents, officers	keyword, automatic truncation	y	y	auto		**Search Companies:** http://sos.state.nv.us/default.asp ***Trademark information and forms (no database):*** http://sos.state.nv.us/comm_rec/trademk/index.htm	co
New Hampshire	company names	automatic truncation	y	n	auto L and R		**Business Name Lookup:** http://199.192.9.86/corporate/default.htm ***Trademark/Servicemark Forms and Laws:*** http://www.nh.gov/sos/corporate/trademarkleader.htm	co
New Jersey	trademarks, company names	browse index, automatic truncation ignores spaces	n	y	auto	ignores spaces	***Search Companies, Trademarks:*** https://accessnet.state.nj.us/GatewayWatchNameSearch.asp	both
New Mexico	company names, they will search trademarks	keyword	y	y	n		***Search Companies:*** http://secure.sos.state.nm.us/ucc/UCCSRCH.HTM ***Trademark information:*** http://www.sos.state.nm.us/trade.htm	co
New York	company names	browse, keyword, partial word (including internal)	y	y	menu		***Search Company names:*** http://wdb.dos.state.ny.us/corp_public/corp_wdb/corp_search_inputs.show ***Trademark forms and instructions:*** http://www.dos.state.ny.us/corp/miscfae.html	co
North Carolina	company names; no TR database on web 11/14/03	browse, keyword. Automatic truncation. Can display report page images.	y	y	auto in "all words"	images	***Search Companies:*** http://www.secretary.state.nc.us/Corporations/ ***TM information:*** http://www.secretary.state.nc.us/trademrk/default.asp	co
North Dakota	company names, trademarks	browse index, keyword, incl. description of mark	y	y	menu	descrip.	***Search Companies and Trademarks:*** http://www.nd.us/sec/entitynamesearch.htm	both
Ohio	trademarks, company names, churches, includes some images	keyword, phonetic. Search "image" or "design" to see examples of images	y	n	menu for sound, sp. stem	images	***Search Companies and Trademarks:*** http://serform2.sos.state.oh.us/pls/porthope/DEV.RPT_BUSINESS_INFORMATION_SQL.SHOW_PARMS	both
Oklahoma	company names: they search trademarks and companies for $5 fee	browse index, keyword, automatic truncation	y	y	auto L and R		***Search Companies:*** http://www.state.ok.us/businessdir/businessdir.php ***TM information:*** http://www.state.ok.us/~sos/functions/trademarks.htm ***"Business Records" search including trade names for $5:*** http://www.sos.ok.us/business/business_records.htm	co
Oregon	company names; TM search for $50 fee	browse, keyword, phonetic, synonyms	y	y	auto, menu	synonyms, sound alike	***Search Companies:*** http://sos-venus.sos.state.or.us:8080/beri_prod/PKG_WEB_NAME_SRCH_INQ_LOGIN ***Forms & fees, including TM information:*** http://www.sos.state.or.us/corporation/forms/index.htm ***Special searches, including TM:*** http://www.sos.state.or.us/corporation/searches/beri_special_searches.htm	co
Pennsylvania	company names, including old names	browse index	n	y	n		***Search Companies:*** https://www.dos.beta.state.pa.us/CorpsApp/CorpsWeb/wfDefault.aspx ***TM information:*** http://www.dos.state.pa.us/corps/mark.htm	co

171

APPENDIX (continued)

STATE	INFORMATION	SEARCH FEATURES	Key	BR	PHR	TRUNC	OTHER	URL	
Puerto Rico	trademarks with images, search companies and trademarks	monthly gazette with TM images; company database is browse index	b	n	n	auto L and R in TM database	images in color	***Avisos Marcas and Nombres Comerciales (monthly notices of new registrations, beginning Sept 2001):*** http://www.estado.gobierno.pr/Avisos_Marcas.htm ***TM Search:*** http://www.estado.gobierno.pr/prpto/MARCAS.ASP ***Corporation register (database):*** http://www.estado.gobierno.pr/CorporacionesOnLine.asp	both
Rhode Island	company names, agents, officers	browse, keyword, right and left truncation or both	y	y	y	L and R %		***Search Companies:*** http://www2.corps.state.ri.us/corporations/new_corp_search/ ***TM information and forms:*** http://155.212.254.78/trademarks.htm	co
South Carolina	company names, agents	browse, keyword, automatic truncation (including internal)	y	y	y	auto	single word or phrase, no AND	***Search Companies:*** http://www.scsos.com/corp_search.htm ***TM information:*** http://www.scsos.com/trademarks.htm	co
South Dakota	They do database search: no web database	no web database						***Companies information:*** http://www.state.sd.us/sos/Corporations/domcorpintofees.htm ***Trademark information:*** http://www.state.sd.us/sos/Trademarks/Trademarks%20Intro%20Page.htm	none
Tennessee	trademarks with images, company names	keyword, automatic truncation, phrases, search "design" or "image" to see images	y	n	n	auto	images	***Search Trademarks:*** http://www.ja.state.tn.us/sos/iets2/ietm/PgTrademarkSearch.jsp ***Search Companies:*** http://www.tennesseeanytime.org/soscorp/	both
Texas	trademarks, company names, agents, officers	requires registration; $1 per search						http://www.sos.state.tx.us/corp/sosda/index.shtml	both-pay
Utah	trademarks including description of images, company names	keyword, automatic truncation, phrases	y	n	y	auto	descrip.	***Search Companies and Trademarks:*** http://www.utah.gov/serv/bna	both
Vermont	trade names, trademarks, company names	browse indexes, company keyword in separate database	y	y	n	auto	5 separate files	***Search Tradenames:*** http://www.sec.state.vt.us/seek/tradseek.htm ***Search Trademarks:*** http://www.sec.state.vt.us/seek/markseek.htm ***Search Corporations:*** http://www.sec.state.vt.us/seek/corpseek.htm ***Search Personal name (officers, etc.):*** http://www.sec.state.vt.us/seek/name.htm ***Trademark and Company Keyword:*** http://www.sec.state.vt.us/seek/keyword.htm	both

State	Search types	Browse method			%, L and R	use L and R % with phrases	Info	
Virginia	trademarks–a few with images; (register to search companies)	browse, keyword (use L and R % if mark is more than 1 word)	y	y			**Search Trademarks:** http://securities.scc.state.va.us/SerfisApp/SerfisCartridge/wbq_tmsm$.startup *Company search (Registration required but no cost):* http://www.state.va.us/scc/division/clk/diracc.htm	both
Washington	company names, agents	keyword	y	n– only if full mark	n	"exact match" = full mark	**Search Companies:** https://www2.wa.gov/sos/cor/search.jsp *Fee schedules, including TM:* http://www.secstate.wa.gov/corps/registration_fee_schedule.aspx?m=undefined **Trademark Information:** http://www.secstate.wa.gov/corporations/registration_fee_schedule.aspx#trademarks	co
West Virginia	trademarks, company names	browse index	n	y	auto on browse		**Search Companies, including registration/reservation names:** http://129.71.220.230/wvcorporations/verifylogon.asp **TM & Trade Name Info:** http://www.wvsos.com/business/filing/otherfilings.htm **Search fee information:** http://www.wvsos.com/business/trademark/formsfees.htm	co
Wisconsin	company names	browse index (alphasort gives more results, use advanced for exact match)	n	y	auto on browse	alphasort ignores some words	**Search Companies:** http://www.wdfi.org/corporations/crispix/ **TM information:** http://badger.state.wi.us/agencies/sos/trade.htm	co
Wyoming	company names	browse index	n	y	auto on browse		**Company and TM search:** http://soswy.state.wy.us/Corp_Search_Main.asp **Bucking horse and rider information:** http://soswy.state.wy.us/bucking/info.htm **Trade Name information:** http://soswy.state.wy.us/corporat/tn.htm **TM info:** http://soswy.state.wy.us/corporat/tm.htm	co

173

Index

Page numbers followed by an n indicate a note.

Acceptable Identification of Goods and Services Manual (USPTO), 139,144
Additional Improvement Patents 1837-1861 (National Archives), 50
Adobe Acrobat Reader, 69
African Regional Property Organization, 95
Alabama
    patent depository library, 52
    trademark database, 168
Alaska
    patent depository library, 52
    trademark database, 168
Alien Property Custodian (APC)
    controversy, 18
    current documents, 18
    information dissemination, 14-16
    licensing programs, 16-18
    patent data, 16-18
    seizure, 7-9
    wartime patents, 5
    World War II, 7-14
American Chemical Society, 15
Annual Report of the Commissioner of Patents, 44
Anthrax scare, drug patents, 6
Arab states
    planned system, 97
    regional patents, 96
Arizona
    patent depository library, 52
    trademark database, 168
Arkansas
    patent depository library, 52
    trademark database, 168

Arnold, T. W., 14
Austria
    regional patents, 92
    Web site, 77

Banqui Agreement (1982), 95
Basic Facts About Trademarks (USPTO), 138,148
Bayer, drug patent suspension, 6
Bayh-Dole Act, 125
Belgium
    licensing protest, 13
    Web site, 10
Biddle, F., 10
Board Information System Index (BISX), 142
Bonneville Power Administration, 15
Borchard, E., 11
Bulgaria, Web site, 77
Business Week (1942), 10
Buttons to Biotech: U.S. Patenting by Women 1977-1996 (USPTO), 50

California
    patent depository library, 52
    trademark database, 168
Canada, drug patents, 6
Caribbean, planned system, 97
Carnegie Library (Pittsburgh), inventor database, 48
Cartagena Agreement, 97
Certification marks, 140
Chemical Abstracts, 34,107
Clark, T. C., 17

Classification and Search Support Information System (CASSIS)
  information files, 51
  patent data, 27,30,35n
  trademark search, 138
CASSIS2
  origins, 158n
  trademark search, 141
Classification Information Files (USPTO), 51
Coca-Cola Company, 24
Collective membership marks, 140
Colorado
  patent depository library, 52
  trademark database, 168
Company name
  state databases, 161-167
  type of access, 163
  Web sites, 161-167
Connecticut
  patent depository library, 52
  trademark database, 168
Copyright, defined, 24
Council for Mutual Economic Assistance, 98
Crowley, L. T., 9
Cyprus, Web site, 77
Czech Republic, Web site, 77

Databases
  company name, 161
  comparison, 112
  downloading, 112
  dual search, 31
  Europe network, 57
  formatting, 112
  historical analysis, 112
  inventors, 48
  numerical type, 31
  patent data, 106-112
  query type, 112
  state trademarks, 161
  text type, 31
  trademarks, 141-144

USPTO vs. CASSIS, 142
  see also Web sites
Delaware
  patent depository library, 52
  trademark database, 168
Denmark, Web site, 77
Deoxyribonucleic acid (DNA), patent database, 106
Delphion, patent data, 109
Design patents, 25
Design Search Code Manual, 144,146
Detroit Public Library, inventor database, 48
Deutschland (submarine),
Dialog
  patent data, 111
  state trademarks, 162
  Web site, 111
Digest of Patents 1790-1839, 46
District of Columbia, patent depository library, 52
Dopbyns, K. W., 49
Drug patents, anthrax scare, 6

eBay, trademark search, 157
Estonia, Web site, 77
Europe
  patent databases, 57
  regional patents, 91-93
European Patent Convention (1973), 58
European Patent Office (EPO)
  application search, 62
  classification systems, 65
  company name search, 63
  current members, 58,91
  database network, 58-61
  full text access, 67-70
  future developments, 79
  gateway, 60
  integrated framework, 82-84
  inventor database, 48
  office sites, 76-79
  online help, 80
  priority system, 66

quick search, 61-63
regional patents, 91-93
results display, 67-70
searches, 70-72
simple text search, 61
Web site, 59
worldwide documents, 68
European Union, regional patents, 91
Eurasian Patent Office, 94

Facsimile Images of United States Patents (USPTO), 51
Federal Deposit Insurance Corporation (FDIC), 9
Finding List for United States Patent, Design, Trademark, Reissue, Label, Print and Plant Patent Numbers (Randall and Watson), 49
Finland, Web site, 79
Florida
    patent depository library, 53
    trademark database, 169
France, Web site, 77
Frequently Asked Questions About Trademarks (USPTO), 138

Genealogical research, patent information, 39-44
General Information Concerning Patents (USPTO), 103
General Motors, 13
Georgia
    patent depository library, 53
    trademark database, 169
Germany
    British blockade, 8
    patent documents, 70
    property seizure, 8
    regional patents, 92
    Web site, 77
Global positioning system (GPS)
    classification, 121
    data retrieval, 120
    defined, 119
    development, 129
    further research, 132
    internationality, 127, 131
    inventors by country, 132
    key players, 125-128
    methodology, 120
    new technology, 124
    notable inventors, 128
    patent data, 102, 117
    search and analysis, 121-125
    technology trends, 117
    types of holders, 126
Goods and services, international classification, 139
Gulf Cooperation Council, 96

Handbook of Experimental Physics (Wien-Hams), 18
Handbook on Organic Chemistry (Beilstein), 17
Harare Protocol (1982), 95
Harwell, K., 139
Havana Agreement (1949), 98
Hawaii
    patent depository library, 53
    trademark database, 169
Hellenic Republic, Web site, 77
Historical and Interesting U.S. Patents in Celebration of Our Nations's Bicentennial (USPTO), 49
Historical Notices of Inventions from Archives of United States (USPTO), 50
Historical research, patent information, 39-44
Hitler, A., 7
Hungary, Web site, 77

Idaho
    patent depository library, 53
    trademark database, 169
Illinois
    patent depository library, 53
    trademark database, 169

Indiana
    patent depository library, 53
    trademark database, 169
Information dissemination, APC
        program, 14-16
Intellectual property, types, 24
Internet
    historic patents, 40
    search methods, 41
    see also Databases; Web sites
Inventions
    classification, 26-29
    historic, 41
    matching problem, 33
    search engines, 43
    1770-1873, 44
    statutory registrations, 25
Inventors
    alphabetical list, 46
    bibliography, 44-51
    databases, 48
    geographic region, 47
    historical, 44-51
    key players, 125
    by subject matter, 44
    trademark search, 156
    women, 50
Iowa
    Inventor database, 48
    patent depository library, 53
    trademark database
Ireland, Web site, 77
Italy, Web site, 77

Japan
    patent data, 68
    searches, 75

Kansas
    patent depository library, 53
    trademark database, 169
Karrh, P., 140
Kelley, R., 140

Kentucky
    patent depository library, 53
    trademark database, 69
Kettering, C., 13
Key word searching vs. classification, 34
Krasner, N. F., 128

Latvia, Web site, 77
Legal Encyclopedia (Nolo Press), 103
LexisNexis
    patent database, 110
    state trademarks, 163
    Web site, 110
Licensing programs, 16-18
Liechenstein, Web site, 77
*Lists of Patents for Inventions and
        Designs 1770-1847*, 46
Lithuania, Web site, 77
Louisiana
    patent depository library, 53
    trademark database, 169
Luxembourg, Web site, 77

Maine, trademark database, 170
Manual of Patent Classification
        (USPTO), 27,103
Manual of Patent Examining Procedure,
        18
Markham, J. E., 16
Maryland
    patent depository library, 53
    trademark database, 170
Massachusetts
    patent depository library, 54
    trademark database, 170
Mein Kampf (Hitler), 7
Miami University, inventor database, 48
Michigan
    patent depository library, 54
    trademark database, 170
Middle East, regional patents, 96
Minnesota
    patent depository library, 54
    trademark database, 170

Mississippi
  patent depository library, 54
  trademark database, 170
Missouri
  patent depository library, 54
  trademark database, 170
Monaco, Web site, 77
Monopoly (game), 24
Montana
  patent depository library, 54
  trademark database, 170
Morgenthau, H., 10

Name and Date Patents 1790-1836, 49
Nation, The (1942), 13
National Archives and Records Administration, 18
National Chemical Exhibition (1942), 14
National Patent Planning Commission, 13
Nebraska
  patent depository library, 54
  trademark database, 170
Netherlands
  licensing protest, 13
  regional patents, 92
  Web site, 77
Nevada
  patent depository library, 54
  trademark database, 171
New American State Papers 1789-1860, 47
New Hampshire
  patent depository library, 54
  trademark database, 171
New Jersey
  patent depository library, 54
  trademark database, 171
New Mexico
  patent depository library, 55
  trademark database, 171
New Republic (1942), 13
New York
  patent depository library, 55
  trademark database, 171

Nice Agreement, 159n
Nichols, M. E., 128
North Carolina
  patent depository library, 55
  trademark database, 171
North Dakota
  patent depository library, 55
  trademark database, 171
Norway, licensing protest, 13

Official Gazette (USPTO), 32, 45
Ohio
  patent depository library, 55
  trademark database, 171
Oklahoma
  patent depository library, 55
  trademark database, 171
Online Computer Library Center, 43
Online European Patent Register, 81
Oregon, trademark database, 171
Organization for Economic Cooperation and Development (OECD), 105
Owner terminology, trademarks, 146

Papers and Abstracts Relating to Early American Inventors (USPTO), 50
Paradise (To Be) Regained (Thoreau), 1
Paris Convention for the Protection of Industrial Property (1883), 66
Patent and Trademark Depository Libraries (PTDL)
  Name and Date Patents, 49
  search steps, 23
  see also Individual states
Patent Cooperation Treaty, 69, 73
Patent Marketing and Information Section, 14
Patent Office Pony, The: A History of the Early Patent Office (Dobyns), 49
Patent searching
  basic steps, 23
  genealogical, 39-44

goal, 40
international, 89-99
methods, 41
tools, 36-38
Patents
abandoned, 17
analysis, 107,121
applications, 102
cartels, 14
catalogs, 14
classification, 26-33
databases, 31,102,106
depository libraries, 52-56
electronic form, 23,51
European, 65
examination, 32,92
further research, 132
genealogical research, 39-44
GPS data, 117
granting, 92
historical research, 39-44
information programs, 16-18
intelligence gathering, 106
international, 65,68,127
Japanese database, 75
key players, 125-128
library network, 15
licensing programs, 16-18
manuals, 15
matching problem, 33
monopolies, 14
nationalizing, 11
new search step, 32
Nordic system, 90
plant, 25
public inspection, 15
retrieval methods, 106-112
special materials, 49
technology assessment, 101,117
vs. trademarks, 24
types, 25
valuation, 103-106
wartime, 5,10,12
World War II, 5-18
see also Regional patents

Pennsylvania
special depository library, 55
trademark database, 171
Plant patents, 25
Poland, Web site, 77
Portugal, Web site, 77
Portable Document Format (PDF), 67,163
Portals, patent search, 83
Priority system, European patents, 66
Professional associations, patent information, 15
Puerto Rico
patent depository library, 55
trademark database, 172

Quick searches, European patents, 61-63

Randall, M., 49
Regional patents, 89-99
basic principle, 90
European patents, 91
international search, 89
offices, 91-98
operational systems, 91
planned systems, 97
South America, 93
Resources for Searching Common Law Trademarks (Harwell), 139
Rhode Island
patent depository library, 56
trademark database, 172
Romania, Web site, 77
Roosevelt, F. D., 9

Search engines
state trademarks, 165
unusual features, 166
Seiko Instruments, 131
Service marks, 140
Silicon Valley, 133
Slovakia, Web site, 77
Slovenia, Web site, 77

Smaller War Plants Corporation, 16
Sneed, M. C., 1
South America, regional patents, 97
South Carolina
  patent depository library, 56
  trademark database, 172
South Dakota, trademark database, 172
Soviet Union
  patent information, 18
  regional patents, 94
  World War II, 13
Spain, Web site, 77
Special Libraries Association, 15
Special materials, patents and
    trademarks, 49
Standard Oil, 13
State trademarks
  access type, 163
  databases, 161-173
  PDF, 163
  search engines, 165
  Web sites, 161-167
Statutory invention registrations, 25
Stone, I. F., 13
Subject Matter Index for Inventions
    (USPTO), 44
Succession of Forest Trees, The
    (Thoreau), 1
Sweden, Web site, 77
Switzerland, Web site, 77

Technology assessment
  development, 104-106
  further research, 132
  intelligence gathering, 106
  internationality, 131
  patent data, 101,117
  retrieval methods, 106-112
Temporary National Economic
    Committee, 14
Tennessee
  patent depository library, 56
  trademark database, 172
Texas
  patent depository library, 56
  trademark database, 172

Thoreau, H. D., 1
Trade secrets, 24
Trading With the Enemy Act (1917), 7
Trademark Applications and Registration
    Retrieval, 142
Trademark Electronic Search System
    (TESS)
  basic information, 138
  databases, 141-144
  word mark fields, 156
Trademark Manual of Examining
    Procedure, 138
Trademark Trial and Appeal Board, 140
Trademarks
  common law, 161
  databases, 141-144
  defined, 140
  depository libraries, 52-56
  field definitions, 151
  manuals, 143
  owner terminology, 146
  vs. patents, 24
  quick searches, 143
  registration, 142,147
  sample queries, 145,148
  search methods, 137
  special materials, 49
  see also State trademarks
Trademarks: More Than Meets the Eye
    (Karrh and Kelley), 140
Trimble Navigation, 129
Turkey, Web site, 77

United Kingdom
  National Patent Office, 78
  Patents Act (1977), 90
  Web site, 77
United States
  European coverage, 81
  patent data, 68,104
  utility patents, 26
United States Patent and Trademark
    Office (USPTO)
  bibliographic database, 109
  databases, 141

resources, 36-38
search steps, 23
trademark search, 138
Web site, 2,108
Utah
    patent depository library, 56
    trademark database, 172
Utility patents, 25

Vermont
    patent depository library, 56
    trademark database, 172
Versailles Peace Conference (1919), 9
Virginia
    patent depository library, 56
    trademark database, 173

Washington
    patent depository library, 56
    trademark database, 173
Watson, E. B., 49
Web sites
    company name, 161
    state trademarks, 161
West Virginia
    patent depository library, 56
    trademark database, 173

Wilson, W., 9
Wisconsin
    patent depository library, 65
    trademark database, 173
Women inventors, 50
Women's Contributions in the Field of Invention (USPTO), 50
World Intellectual Property Organization, 73
World Patents Index (Derwent), 106
World Trade Organization (WTO), 99
World War I, property seizure, 7-9
World War II
    enemy information, 5
    patent data, 5-18
    property seizure, 9-14
Writings of Henry David Thoreau, The (1906), 1
Wyoming
    inventor database, 49
    patent depository library, 56
    trademark database, 173

X-search, trademarks, 138

Yale University, 11

SPECIAL 25%-OFF DISCOUNT!

Order a copy of this book with this form or online at:
http://www.haworthpress.com/store/product.asp?sku=5053
Use Sale Code BOF25 in the online bookshop to receive 25% off!

Patent and Trademark Information
Uses and Perspectives

____ in softbound at $18.71 (regularly $24.95) (ISBN: 0-7890-0440-2)
____ in hardbound at $26.21 (regularly $34.95) (ISBN: 0-7890-0425-9)

COST OF BOOKS ____
Outside USA/ Canada/
Mexico: Add 20% ____

POSTAGE & HANDLING ____
(US: $4.00 for first book & $1.50
for each additional book)
Outside US: $5.00 for first book
& $2.00 for each additional book)

SUBTOTAL ____
in Canada: add 7% GST ____

STATE TAX ____
(CA, IN, MIN, NY, OH, & SD residents
please add appropriate local sales tax

FINAL TOTAL ____
(if paying in Canadian funds, convert
using the current exchange rate,
UNESCO coupons welcome)

❏ BILL ME LATER: ($5 service charge will be added)
(Bill-me option is good on US/Canada/
Mexico orders only; not good to jobbers,
wholesalers, or subscription agencies.)

❏ Signature ____

❏ Payment Enclosed: $ ____

❏ PLEASE CHARGE TO MY CREDIT CARD:
❏ Visa ❏ MasterCard ❏ AmEx ❏ Discover
❏ Diner's Club ❏ Eurocard ❏ JCB

Account # ____

Exp Date ____

Signature ____
(Prices in US dollars and subject to change without notice.)

PLEASE PRINT ALL INFORMATION OR ATTACH YOUR BUSINESS CARD

| Name |
| Address |
| City | State/Province | Zip/Postal Code |
| Country |
| Tel | Fax |
| E-Mail |

May we use your e-mail address for confirmations and other types of information? ❏Yes ❏ No
We appreciate receiving your e-mail address. Haworth would like to e-mail special discount offers to you, as a preferred customer. **We will never share, rent, or exchange your e-mail address.** We regard such actions as an invasion of your privacy.

Order From Your Local Bookstore or Directly From
The Haworth Press, Inc.
10 Alice Street, Binghamton, New York 13904-1580 • USA
Call Our toll-free number (1-800-429-6784) / Outside US/Canada: (607) 722-5857
Fax: 1-800-895-0582 / Outside US/Canada: (607) 771-0012
E-Mail your order to us: Orders@haworthpress.com

Please Photocopy this form for your personal use.
www.HaworthPress.com

BOF03

DURANGO PUBLIC LIBRARY
DURANGO, COLORADO 81301